Herbarium

Herbarium

The Quest to Preserve & Classify the World's Plants

Barbara M. Thiers

TIMBER PRESS • PORTLAND, OREGON

For my parents, Ellen J. and Harry D. Thiers,
and my mentor, Patricia K. Holmgren

Maps by Elizabeth Gjieli, New York Botanical
Garden GIS Laboratory.

Photo credits are on page 259.

Published in 2020 by Timber Press, Inc.
The Haseltine Building
133 S.W. Second Avenue, Suite 450
Portland, Oregon 97204-3527
timberpress.com

Printed in China
Jacket and text design by Adrianna Sutton

ISBN 978-1-60469-930-2

Catalog records for this book are available from
the Library of Congress and the British Library.

MIX
Paper from
responsible sources
FSC
www.fsc.org   FSC® C104723

Genus *Claudea*

Species *elegans Lamour.*

1656

1656. *FLORA LUDOVICIANA.*

Ascochyta asteroides, E & E
6.7×10 μ
Perith. very thin, no asci apparent, sporidia
one-two guttulate (some septate?)
Hyphop. on indefinite spots of lvs of
Magnolia grandiflora

LEGIT A. B. LANGLOIS, 28 I 1889
St Martinsville P. O., La

Asterinula Langloisii, E & E
Can not be Ascochyta
on account of the superficial
perithecia

= Septothyrella Langloisii (E. + E.) Sacc.
Sacc. Syll. Fung. X. 426.

Contents

!McDaniel 1966

Preface

I've been around herbaria all my life. My father was founder and curator for 40 years of the herbarium at San Francisco State University (now the Harry D. Thiers Herbarium). In addition to teaching, he poured his considerable curiosity and energy into documenting the fungi, primarily, but also the bryophytes (i.e., mosses and liverworts) and lichens of California. My childhood weekends were spent either on field trips or in the herbarium, where I could earn some pocket money by helping my dad with his specimens. I have fond memories of Sunday afternoons spent gluing specimen labels to boxes as my dad typed them while we listened to Giants baseball games on the radio. As a teenager, I wanted nothing to do with the herbarium, or my parents for that matter, but eventually, in the middle of my college years, I found my way back. I discovered that I too wanted to participate in the great adventure of documenting plant and fungal life—and to join my dad and his passionate, sometimes eccentric band of students and colleagues who derived such joy from this work.

◀ An herbarium specimen of
Sarracenia purpurea (common pitcherplant).

I arrived at the New York Botanical Garden in 1981 to start a one-year postdoctoral fellowship in the William and Lynda Steere Herbarium. Through my time in my dad's herbarium, I understood the management issues in a small collection built specifically for the research and teaching needs of a contemporary scientist. I was youthfully confident that these skills would transfer well to one of the largest and most active herbaria in the world. It did not take long for me to realize how little I knew.

My first project at the Garden was to organize and index the collections of the William Mitten herbarium. Mitten was a 19th-century English apothecary and self-taught scientist who had obtained bryophyte specimens from essentially all the famous European exploring expeditions. For this project, I handled specimens collected by the likes of Alexander von Humboldt, Joseph Banks, Charles Darwin, William and Joseph Hooker, and many collectors more obscure but equally fascinating. The work was thrilling but frustrating: often these historic specimens lacked adequate documentation—a scribble in terrible handwriting might be all that accompanied a historic type specimen, one that has been the basis for a scientific name in use for a hundred years, and often there was no ready source for the information needed to fill in the missing details. From that experience I developed an enduring interest in botanical exploration and historical collections. I loved to read about the adventure but was happiest when I was able to learn more about how collectors approached the less exciting but ultimately more important work of preserving and studying the specimens they collected.

I managed to stay on in a permanent capacity at the Garden after that first year, eventually working my way up through the ranks of herbarium administration. Around the time I became director of the herbarium, I was asked to participate in a workshop, sponsored by the National Science Foundation, about the digitization of herbarium specimens. The facilitator began the workshop by asking the 15 or so participants, most of whom had worked in herbaria for years, to create a timeline of the important dates in the history of herbaria. All of us were stumped. We couldn't agree on any milestones. I was dismayed by yet another example of how ill-informed I was about my own supposed area of expertise but a bit relieved that my colleagues were in the same boat. How odd that even though herbaria are absolutely fundamental to our knowledge of plant and fungal diversity, formal botanical training in the mid- to late 20th century included so little in the way of background about them!

Over the years I've continued to try to fill the gaps in my herbarium knowledge, and to talk about the importance of herbaria to a variety of audiences. When Timber Press approached me, I was glad for the opportunity of putting some of what I have learned into book form. I hope that the result will be a useful introduction to herbaria for natural history enthusiasts in general but also for colleagues to share with students, new herbarium staff, or maybe even deans or other institutional leaders. I also want to demonstrate, through this book, the wide range of circumstances under which herbarium specimens have been gathered and handled after collection, which accounts for the inconsistency of the data that accompany them. Understanding the journeys of specimens may help users of data derived from herbaria to be aware of both their strengths and limitations.

I also hope that this book will engender an appreciation for institutions that have made the commitment to preserve specimens of plants and fungi in perpetuity. At a time when we seem to be bombarded daily by negative aspects of human nature, herbaria highlight one of our better human impulses: to save things for the future, not just for ourselves but for generations to come. It is true that many times plant and fungal collection has been fueled by imperialism. The impulse to kill these organisms just to store them in cabinets may seem to derive from some base hoarding instinct. However, investing time and resources to care for these irreplaceable representatives of past plant and fungal life and sharing them freely with anyone wanting to study them is both generous and wise. We should celebrate the botanic gardens, museums, universities, and government agencies worldwide that year after year commit their resources to this noble endeavor for the benefit of us all. As much as our modern lives tend to separate us from the rest of earth's biodiversity, we cannot exist without it, and these preserved organisms give us information about our world and clues to its future that we cannot learn any other way.

The hardest part of writing this book has been deciding what not to include. In order not to tax the patience of my readers and editors, I could accommodate only a small sample of the stories about the sometimes odd but loveable people who have helped build the world's herbaria. Also, I could not focus equal attention on herbaria worldwide—a realization that saddened me greatly. One of the great joys of my career has been to get to know curators of herbaria all over the world, through my travels and my years as editor of *Index Herbariorum* (sweetgum.nybg.org/science/ih), an online registry of and guide to the world's herbaria. Many of these colleagues, especially those in developing countries, have shown amazing ingenuity and dedication to building and maintaining their herbaria, and are as worthy of attention as anyone included in this book. Because of limitations of space and my own knowledge and foreign language expertise, I have focused most of my attention on collecting and herbaria in Europe, where the tradition originated, and in the United States, although I do include brief histories of herbarium development in four other selected countries. More information about the herbaria mentioned is available at the end of the book. I hope my colleagues in other countries will write the story of their herbaria one day.

A

B

C

US
152

Plants of MARYLAND, U.S.A.

Lupinus perennis Linnaeus

ANNE ARUNDEL CO.: Town of Odenton, power line
right of way parallel to MD Rt. 32 near the intersection
of MD Rt. 175 and MD Rt. 170, growing in full sun in
dry sandy soil, scattered.
39° 05' 09.26" N, 76° 41' 14.65" W.

28 April 2011

Wayne D. Longbottom   14,791
Gary Van Velsir

D

The Origin *of* Herbaria

For the past almost six centuries, scientists have been documenting the plants and fungi of the world through herbaria. The basic preparation of the specimens that are housed in an herbarium has changed relatively little over time. But the invention of this simple technology was a key innovation in transforming the study of these organisms from a minor subdiscipline of medicine into an independent scientific endeavor. The herbarium made it possible for scientists to characterize the plants and fungi that grow in faraway places and to understand their diversity on a global scale. Our wealth of specimens today, so carefully preserved through the centuries, is a unique source of data that helps us understand how the world's vegetation has changed over time and predict how it will change in the future.

◄ An herbarium specimen of *Lupinus perennis* (sundial lupine).
A: The pressed plant, with roots, stems, leaves, and flowers. B: A separate leaf, showing both the upper and lower surfaces of the leaflets. C: The specimen label, which indicates the name of the plant, where and when it was collected, and by whom. D: A fragment packet, in which any parts that come off of the specimen can be stored. E: The barcode number, a unique identifier for specimens that have been digitized. F: A stamp indicating the herbarium that owns the specimen, in this case, the William and Lynda Steere Herbarium of the New York Botanical Garden.

Luca Ghini and the
Origin of the Herbarium

Evidence suggests that Luca Ghini, a physician and highly esteemed professor, created the first herbarium. The epitome of a Renaissance scholar and teacher, Ghini was born around 1490 near Bologna. His family was part of a growing phenomenon in northern Italy—a middle class composed of merchants and professionals who had the means and time for intellectual pursuits. Forward-looking universities in regional cities such as Pisa, Padua, and Bologna encouraged independent thought, and a continual influx of new ideas and goods came to the region via travelers and traders en route between northern Europe, the Mediterranean region, and the Middle East. Because of the favorable economic and intellectual climate, the Renaissance is said to have come to northern Italy half a century earlier than elsewhere in Europe.

Ghini gained his medical degree from the University of Bologna in 1527. In addition to his medical studies, he had a broad range of interests, including local flora, fauna, and minerals, which he studied on excursions throughout northern and central Italy. The University of Bologna hired him as a lecturer in medicine after he graduated, and a few years later he introduced a new course to the medical curriculum, one devoted entirely to plants, or "simples," as plants were often called at the time. This appears to have been the first such course in an Italian university, perhaps in all Europe. Previously, plants were covered in the university curriculum only as part of pharmacognosy, and such courses mostly relied on information that was centuries old.

Renaissance scholars reevaluated existing knowledge, most of which derived from Greek and Roman sources. In doing so they revealed the limits of classical knowledge and began new inquiries. Most of the knowledge about plants up until Ghini's time came from the soldier and physician Dioscorides, who was born in the 1st century CE (Common Era) in Asia Minor. In his work Dioscorides traveled extensively throughout the Mediterranean region, which gave him the opportunity to study and use a wide range of plants. He summarized the characteristics and medical properties of more than 500 plants in his *De Materia Medica* (or *Peri hules iatrikēs*, "on medical material," in his native Greek), published around 65 CE.

Luca Ghini, from a painting in the Biblioteca Comunale, Imola, Italy. This is a 19th-century copy of an earlier work, which has been lost. It is the only known painting of him.

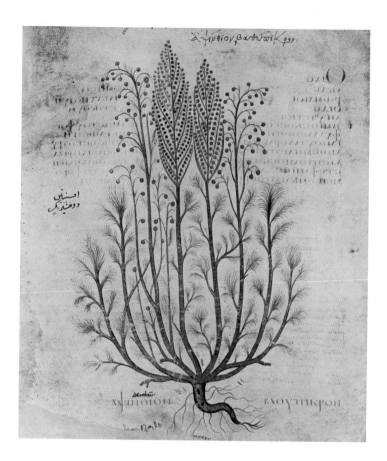

Artemisia absinthium (wormwood), from the Vienna Dioscorides (*Codex Aniciae Julianae picturis Illustratus, nunc Vindobonensis*), an early 6th-century Byzantine Greek illuminated manuscript of *De Materia Medica*.

To make sure that his students gained a fuller view of plant diversity, Ghini supplemented his lectures with living plants and encouraged students to make their own observations. Some of the plants he collected in the surrounding countryside; others came from correspondents from as far away as Crete and Syria. Ghini's concept of teaching through direct observation, rather than depending on classical texts and illustrations, had a direct parallel in other teaching practices of the time, such as the dissection of cadavers to study human anatomy. His course was very popular with students, and within a few years, the subject of botany was a core element of the curriculum, with Ghini as chair of botany. He attracted a circle of formal and informal students who would later spread his ideas and innovations throughout Italy and elsewhere in Europe.

Early in the 17 years Ghini spent teaching his course on plants at Bologna, he had the idea for an innovation that would allow his students to study the important features of plants even in winter when the plants were dead or dormant. This was the herbarium, originally called a Hortus Hiemalis (winter garden) or Hortus Siccus (dry garden). Preparing an herbarium specimen involved placing a freshly gathered plant between sheets of paper in a somewhat naturalistic pose, and then applying pressure to flatten the plant and remove moisture from it. When the specimen was completely dry, it was glued onto the page of a blank book, and the page might then be annotated with a note about the plant's name, its distinctive features, its medicinal properties, or where it was gathered. If handled carefully and kept protected from moisture, insects, and light, a dried plant specimen could be preserved in this manner indefinitely.

The key features of a plant could be observed in a well-prepared herbarium specimen far more easily than in the books available in the early 16th century. Methods of illustration at that time, except by master artists, were not sophisticated enough to capture the level of detail in plant features needed for precise identification. For example, the illustrations in the *Grete Herball* by Richard Banckes, the

▲ *Grete Herball* woodcut of an unidentified plant.

◄ *Tuft of Cowslips* by Albrecht Dürer, 1526.

first English-language herbal, published in 1529, show recognizable plant parts (e.g., stems, roots, and leaves) but certainly could not be used to definitively distinguish one species from another. It would be many years until illustrations of the quality of Albrecht Dürer's *Primula* would be available to the average botany student.

Despite his success at Bologna in creating both a plant study curriculum and the herbarium, Ghini was apparently not completely satisfied with the tools he had created for teaching botany—he also wanted to grow a wide range of native and exotic plants for his students to study. In 1544 he took a position at the University of Pisa, which afforded him the opportunity to not only continue teaching about plants but also to create a living resource. The resulting Hortus Simplicium (later the Orto Botanico) was the first known botanic garden. Although it has been moved several times, it still exists, at 5 Luca Ghini Way, very close to the Tower of Pisa. Ghini obtained support for this project from Cosimo de' Medici (1519–1574), the

The Botanical Garden of Pisa (Orto Botanico di Pisa) today.

duke of Tuscany. The hiring of Ghini and the development of the Orto Botanico were part of a larger effort by the duke to enhance science and the arts in Tuscany. A year after the Pisa garden was established, the University of Padua created a similar one. The supervision and development of the Padua garden, which thrives to this day on its original site, was the work of Ghini's student Luigi Anguillara (c.1512–1570).

Ghini spent 11 years in Pisa, continuing to train a wide range of influential scientists. He returned to Bologna a year before his death in 1556. Ghini was an effective, charismatic, and generous teacher who often maintained a lifelong relationship with his former students. When he died, they exchanged letters expressing great sadness at his passing, and some of his former students raised funds to support the education of Ghini's son and for dowries for his daughters.

Many of Ghini's students became well-known pioneers in botany. In addition to his development of the garden at the University of Padua, Anguillara produced *Semplici* (1549–60), a book series with plant descriptions so precise that the species can be recognized just from his text. Ulisse Aldrovandi (1522–1605) succeeded Ghini at the University of Bologna, where he established a botanic garden and assembled an herbarium of more than 7,000 specimens, the largest of his time; the monotypic genus *Aldrovanda* is named for him. When Ghini left the University of Pisa, his position went to his student Andrea Cesalpino (1519–1603), for whom the genus *Caesalpinia*, in the legume family, is named. Cesalpino is most famous for the insightful classification system he introduced in his *De Plantis Libri* (1583). In this work he groups plants based on their structural features rather than on their medicinal properties or alphabetical order, as had usually been the practice in previous works.

Early in his career Ghini apparently planned to publish a reevaluation of the plants treated in the *Materia Medica* of Dioscorides. However at some point he decided to turn this work over to his student Pietro Andrea Mattioli (1501–1577),

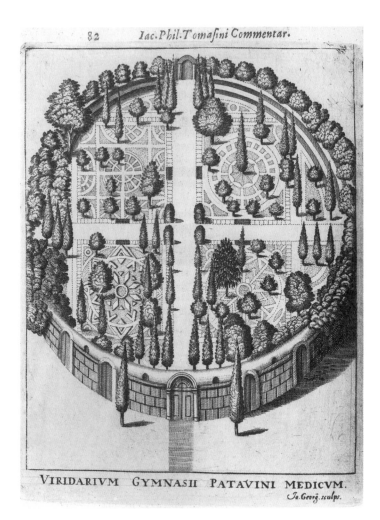

Engraving of the Padua Botanical Garden by Giacomo Tomasini in his guide to the garden, *Gymnasium Patavinum Jacobi Philippi Tomasini* (1654).

▲▲ An engraving of Aldrovandi's cabinet of curiosities. From David Teniers the Younger's *Theatrum Pictorium* (1660).

▲ The aquatic insectivorous *Aldrovanda vesiculosa* (waterwheel plant), the sole member of the genus named for Ulisse Aldrovandi.

◀ Portrait of Andrea Cesalpino by Battista Ricci, held in the Rettorato of the University of Pisa.

▶ *Caesalpinia pulcherrima* (poinciana), named for Andrea Cesalpino. Plate 5, from *Familiar Indian Flowers* (1878) by Lena Lowis.

who published his *Commentarii in sex libros Pedacii Dioscoridis Anazarbei de Materia Medica* in 1544. Ghini not only supplied plants for Mattioli's consideration but also provided suggestions on the manuscript. The *Commentarii* was a monumental work that was published in many editions and translated into Latin, French, Czech, and German. Illustrated by handsome woodcuts, it contained descriptions of some plants that were not in Dioscorides' *Materia Medica* and were of no known medicinal value. Thus, this publication marks a transition in botanical writings from one focused purely on the medicinal properties of plants to one that considered plants for their own intrinsic interest. Mattioli is eponymized in *Matthiola*, a genus in the mustard family which contains *M. incana* (stock), a popular garden plant.

Ghini's herbarium has not survived, and he did not leave any writings asserting that he was responsible for creating this research tool. Indeed, the act of pressing a plant in paper seems so natural and obvious that is hard to imagine it had a single origin. However, he and his students discuss his collection of dried plants in their correspondence, and the herbaria of several of his students still exist. The botanical section of the Natural History Museum in Florence holds the herbaria of two Ghini students: a collection of approximately 200 specimens prepared by Michele Merini, who was a priest in Lucca, and a larger collection, numbering some 700 specimens, of Cesalpino. Aldrovandi's herbarium is maintained in the herbarium of the University of Bologna. Many believe that Ghini's specimens are indeed part of the Aldrovandi collection, but if so, these are not distinguished from those collected by Aldrovandi or others in any way. Guillaume Rondelet (1507–1566), a French physician, ichthyologist, and botanist, studied with Ghini in Pisa for a time and brought the technique of herbarium preparation to France when he became a professor at the University of Montpellier. Through his students, the method spread to Germany, Switzerland, and the Netherlands.

Two other early herbaria of note with a connection to Ghini include the herbaria of Francesco Petrollini, a physician and plant collector from Viterbo, Italy, and Leonhard Rauwolf (1535–1596), a German botanist and one of the earliest explorers of western Asia and Turkey, who studied with Rondelet in Montpellier. Little is known about Petrollini, except that he was a correspondent of Ghini and Aldrovandi, but he is likely the creator of two 16th-century herbaria that still exist today. One is an untitled two-volume set now deposited in the Biblioteca Angelica in Rome, and the other is the *En Tibi* herbarium, now at the Naturalis Biodiversity Center in the

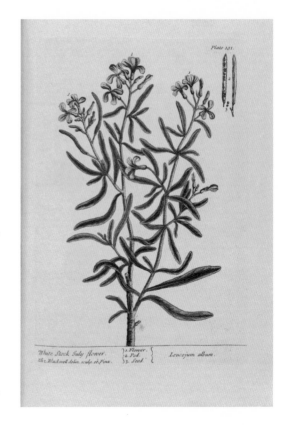

Matthiola incana from Elizabeth Blackwell's *A Curious Herbal*, vol. 1, plate 181. Blackwell illustrated plants from the Chelsea Physic Garden in London for this book, creating the drawings, engravings, and coloring for the images. She used the proceeds from the book to secure her husband's release from debtor's prison. He repaid the favor by leaving her to live in Sweden, where he was later hanged for treason!

Portrait of Pietro Andrea Mattioli, 1533, held in Genoa's Musei di Strada Nuova–Palazzo Rosso.

Netherlands. *En Tibi*, whose full name translates to "Here for You a Smiling Garden of Everlasting Flowers," is a large, beautifully preserved volume meant to include all known types of plants, probably created for sale or as a gift to a wealthy patron. Rauwolf's herbarium consists of four volumes of preserved plant specimens, the fourth of which, prepared with extra care, contains his Middle Eastern collections. The pages of this volume have an edging of thicker, colored paper with a faux wood grain, of a type used as wallpaper to cover beams and ceilings. This thicker border raised the space between the pages of the book, creating extra room for the dried plant material. Both the Rauwolf and *En Tibi* herbaria were at one time owned by Hapsburg Emperor Rudolf II but were stolen from him by the troops of Maximilian of Bavaria in 1620. A short time later, the Swedes looted Maximilian's treasures during the Thirty Years War (1618–48), and the two herbaria came into the hands of the scientifically inclined Queen Christina of Sweden. She later gave the volumes to her Dutch librarian, and the University of Leiden purchased them after his death in 1689.

In 1603, Flemish physician and botanist Adriaan van den Spiegel (1578–1625) published *Isagoges in Rem Herbariam*, giving detailed instructions on how to prepare an herbarium. The instructions describe a process very similar to that used today, although the recipe for glue used to adhere specimens to paper is different: van den Spiegel's recipe called for bull's ears to be boiled with aloe, a piece of steel, and powdered cloves! Today's herbaria generally use a water-based glue, although some, such as the herbarium of the Conservatory and Botanical Garden of Geneva in Switzerland, use straight pins to hold the specimen to the paper.

A variation on the herbaria created by Ghini and his disciples is the so-called *Herbarium Vivum* of Hieronymus Harder (1523–1607), a German physician and teacher with no known connection to Ghini. The *Herbarium Vivum* is a book of dried pressed plants in which the specimen is augmented with a drawing or painting. Sometimes the drawn portion represents missing parts of the plant, such as roots or flowers; in other cases, the habitat for the plant is indicated by a painting of the land surface and perhaps other vegetation at its base. Harder produced 12 volumes of such plant/drawing hybrids, 11 of which are stored in various European institutions. The *Herbarium Vivum*–style of specimen preservation did not catch on, probably because of the labor and skill involved in the painted portion, but two other examples still exist: one created by Johannes Harder (1563–1606), the son of Hieronymus, and another by Henrik Bernard Oldenland (c.1663–c.1697), a German-born South African physician, botanist, painter, and land surveyor.

A beautifully prepared specimen of a jack-in-the-pulpit, *Eminium rauwolffii*, from the fourth volume of Leonhard Rauwolf's herbarium, the specimens for which were collected between 1573 and 1576, on Rauwolf's trip to western Asia and Turkey.

A page from *En Tibi*, showing, among other specimens, a pressed tomato, *Solanum lycopersicum*, including the fruit, from which the pulp has been removed.

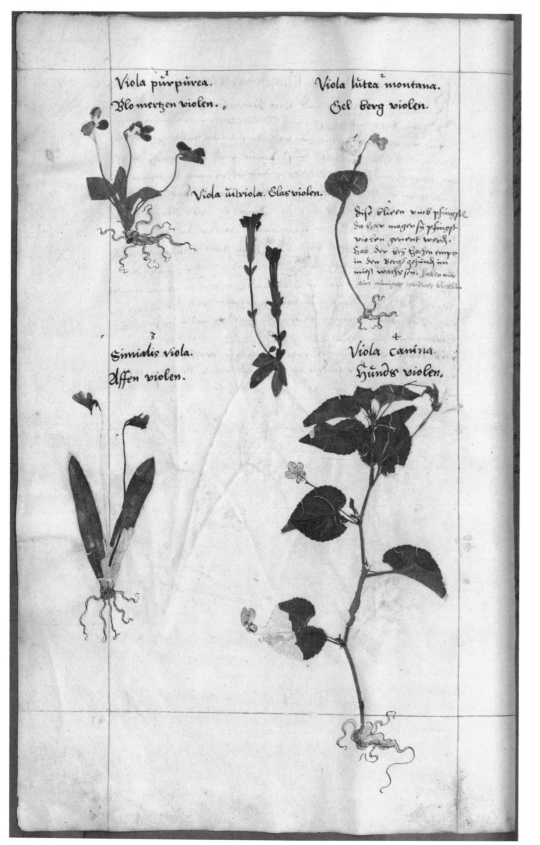

Viola purpúrea.
Blo mertzen violen.

Viola lútea montana.
Gel berg violen.

Viola ûitriola. Elas violen.

Simialis viola.
Affen violen.

Viola canina.
Hünds violen.

Specimen of violets (*Viola*)
from Hieronymus Harder's
Herbarium Vivum.

Origins of Cryptogamic Collections

The earliest herbaria focused on the dominant group of plants on earth, the vascular plants. United by the presence of specialized conductive cells to transport water and nutrients through the plant (a "vascular" system), this group includes ferns and their relatives (clubmosses and horsetails), the gymnosperms (conifers and related groups), and the flowering plants. However, herbaria also hold preserved specimens of algae, bryophytes, and fungi—or, collectively, cryptogams, an archaic but useful term (from the Greek, *kryptos* "hidden," *gameein*, "to marry") that refers to the lack of visible reproductive structures, such as a flower, in these organisms. Some would include ferns in this group as well, because they also reproduce by spores, although since they have a vascular system, their structure is more like flowering plants. Early herbaria contained only a small number of cryptogams, and those that were included were usually the larger, more conspicuous species. The Rauwolf herbarium contains a single bryophyte, *Conocephalum conicum*, a large and extremely common species of liverwort. *En Tibi* contains four mosses and two lichens, all common and conspicuous European species. Aldrovandi's herbarium at the University of Bologna contains 22 specimens of algae, bryophytes, and lichens. Today most herbaria include at least some representatives of cryptogamic groups, and some herbaria specialize in them.

▼ *Codium tomentosum*, a marine green alga.

▼▼ *Sphagnum fallax*, a moss found in bogs. Sphagnum is the principal component of peat.

ALGAE is an informal term for organisms that are photosynthetic and usually aquatic. They may range from single- to multi-celled, and they lack true roots, stems, and leaves. Algae have a wide range of reproductive strategies but do not produce flowers and fruits. Instead, they reproduce by sexual or asexual spores. Examples of algae are diatoms, phytoplankton, "pond scum," and seaweeds. There are an estimated 72,000 species of algae.

BRYOPHYTE is an umbrella term relating to three lineages of plants, including mosses (Bryophyta), hepatics or liverworts (Marchantiophyta), and hornworts (Anthocerotophyta). All are small, terrestrial, photosynthetic organisms that may grow erect, appressed to soil or

rock, or pendent from tree branches. Some bryophytes resemble small leafy plants with distinct leaf and stem-like structures; others are flattened and ribbon-shaped without distinct leaf or stem-like structures. They reproduce by spores that are usually borne on stalks that extend above the level of the plant. There are an estimated 16,000 species of bryophytes worldwide.

FUNGI, now recognized as a separate kingdom, are a very diverse group of organisms that includes yeasts, molds, lichens, mushrooms, polypores, and puffballs. Herbarium collections of fungi also usually contain slime molds or myxomycetes, although these are no longer considered to be fungi. Fungi are actually more closely related to animals than plants. Fungi are heterotrophic, which means they absorb their food rather than creating it through photosynthesis. Most fungi are composed of a network of hyphae (narrow, tubular

▲▲ *Riccia sorocarpa*, a thalloid liverwort.

▲ *Cookeina tricholoma*, an ascomycete fungus. Spores are produced in cells lining the inside of the cups.

branching filaments), which may be loosely interwoven, as in a mold, or tightly fused together, as in a mushroom. The cell walls of hyphae are made of chitin, the same substance that makes up the exoskeleton of insects. Fungi reproduce by spores. Estimates of the number of species of fungi vary, with some as high as 5 million.

For a variety of reasons, 16th-century botanists largely ignored cryptogams. Some of these organisms, like the mushrooms, appear only sporadically. Both mushrooms and algae may decay quickly upon collection and are difficult to preserve. In all the cryptogamic groups, the features that are key for understanding their life histories are too small to view with the human eye alone, which inhibited an understanding of their relationship to other organisms. New methods in glass-making and lens-grinding developed around 1600 made it possible for telescopes and microscopes to become scientific instruments. In 1610, in his *Sidereus Nuncius* (The Starry Messenger), Galileo reported seeing the mountains of the moon through his telescope; his work created controversy but also great interest in the idea of using lenses to see beyond what is visible to the naked eye. The microscope was technically a more difficult instrument to develop than the telescope, and early versions, such as that developed by the Dutch father and son spectacle-makers Hans

and Zacharias Janssen, had a magnification of only about 9×. The microscope developed by Anton van Leeuwenhoek, however, achieved a magnification of 270× and was used by English scientist Robert Hooke in his *Micrographia* (1665), a book of drawings made through the microscope, including one of the cellular structure of cork (the bark of the cork oak, *Quercus suber*). The ability to examine the internal structure of plants provided by the microscope played an important role in advancing knowledge of plant diversity. Scientists such as Nehemiah Grew (1641–1712) and Marcello Malpighi (1628–1694) carried out extensive studies on the comparative anatomy and morphology of plants, which, although focused on vascular plants, served to highlight how they differ from non-vascular plants. John Ray (1627–1705), an English naturalist and theologian, transformed Cesalpino's vision of grouping plants based on natural affinities into a fully realized classification system that was enhanced by the studies of Grew and Malpighi as well as his own observations. In his *Historia Plantarum* (3 volumes, 1686–

▲▲ *Phylloporus rhodoxanthus.* The spores of this mushroom are produced externally from cells lining the surface of the gills, the plate-like structures under the cap.

▲ *Cladonia verticillata*, a lichen. Lichens are fungi that form a symbiotic association with an alga. The spores are produced from the apothecia, the brown structures atop the gray-green stalks. Lichens grow mainly on soil, trees, or rocks.

1704), Ray pioneered the use of the term "cryptogamia" as a unit of classification, which he divided into four orders: filices (ferns and related groups), musci, algae (which included lichens and liverworts), and fungi.

The Jardin du Roi (Royal Garden), founded in Paris in 1635, was one of the earliest centers for the study of cryptogamic plants. Joseph Pitton de Tournefort (1656–1708) studied the origin of spores in ferns, bryophytes, and fungi, although he did not understand how a single-celled spore differed from a seed, which contains a multicellular embryo and its endosperm, or food supply, and a protective coat. René A. F. de Réaumur (1683–1757) was the first to describe the reproductive structures in the marine alga *Fucus* (rockweed), and Jean Marchant (1650–1738) in 1713 first described the hepatic genus *Marchantia*, which he named for his father, Nicolas, who was also a botanist. Marchant made highly accurate drawings of the structure of the plant but still misinterpreted the spores as a very small type of seed.

The botanic garden in Florence (Orto Botanico di Firenze) was also a focal point of the study of cryptogams, led by Pier Antonio Micheli (1679–1737). Micheli was the first to describe reproduction in fungi, and thus he is often considered the founder of the study of mycology (the study of fungi); he is also known for his pioneering studies of bryophytes and lichens. His *Nova Plantarum Genera* (1729), with 73 pages of illustrations, was largely devoted to cryptogams; in it, Micheli described about 1,900 plants, including 900 fungi and lichens. His student Giovanni Targioni Tozzetti (1712–1783) succeeded him as director of the Orto Botanico, and the Micheli-Targioni herbarium (now part of the herbarium at the Natural History Museum in Florence) contains about 19,000 specimens. This herbarium is very important as a source of early specimens of cryptogamic groups, although many collections of algae and fungi have not survived. In some cases, the fungi are represented by Micheli's illustrations. The specimens that do still exist can be difficult to interpret because they are glued on small pieces of paper with very little information about the name or where the specimen was collected.

Never a wealthy man, Micheli depended on subscribers to support the publication of the three volumes of his *Nova Plantarum Genera*. Among these subscribers was William Sherard (1659–1728), who established the herbarium at Oxford University. When Sherard retired, he chose Johann Dillenius (1684–1747) to replace him. Dillenius, born in Darmstadt, Germany, became the first Sherardian Professor of Botany in 1734. Influenced by Micheli, Dillenius followed in his footsteps as a cryptogam specialist, although he had less of an understanding of the importance of microscopic characters than Micheli, causing him to misinterpret the functions of features such as the sporophyte of bryophytes, which he considered to be a flower.

▲ Etching of scientists at work in the laboratory of the Royal Garden in Paris by Sebastien Leclerc (1637–1714).

▶ Plate 86 from Micheli's *Nova Plantarum Genera*, published in Florence in 1729.

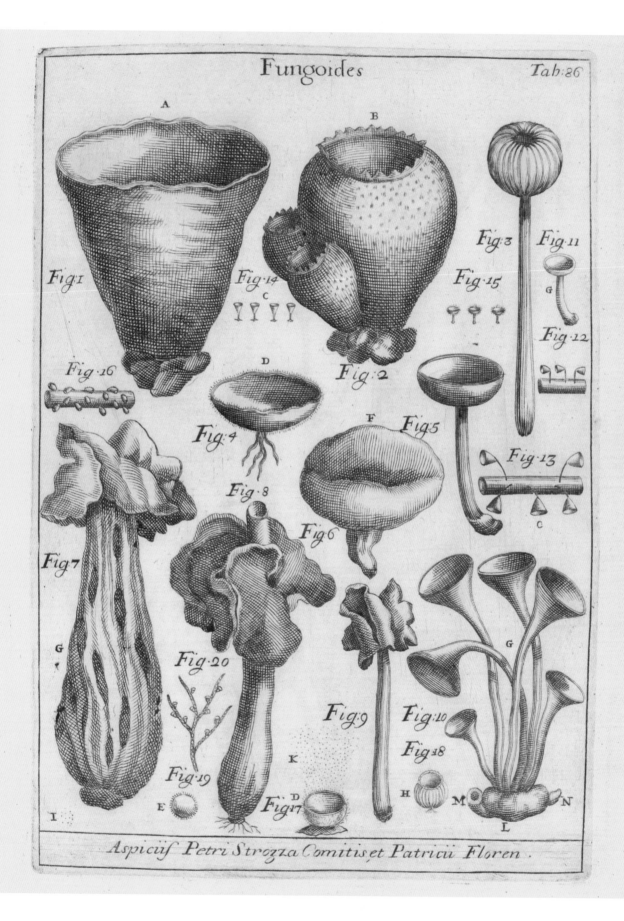

Aspiciif Petri Strozza Comitis et Patricii Floren.

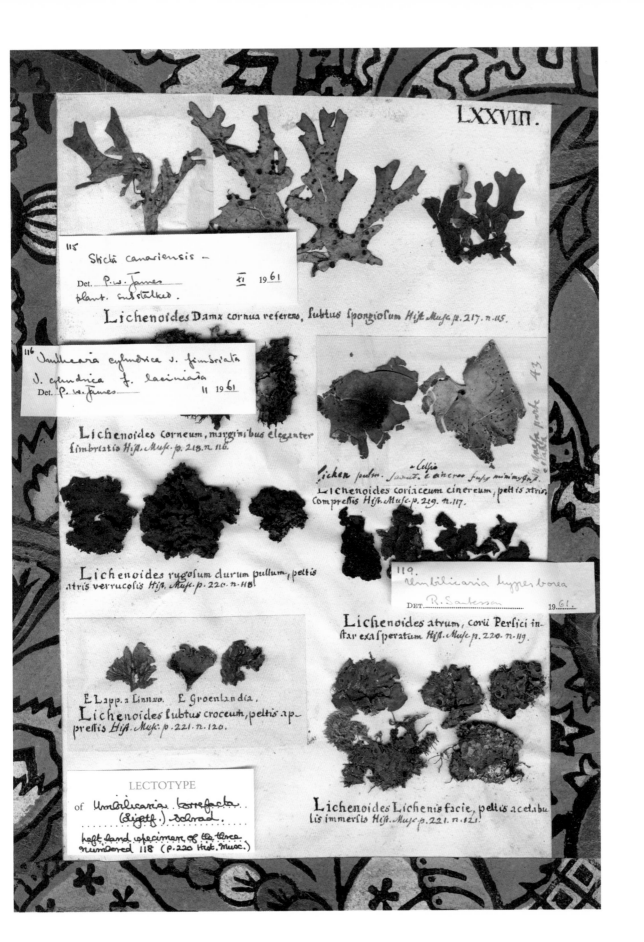

LXXVIII.

115

Sticta canariensis —

Det. P.W. James   xi 19 61

plant. substalked.

Lichenoides Damæ cornua referens, subtus spongiosum Hist. Musc. p. 217. n. 115.

116 Imbicaria cylindrica v. fimbriata

I. cylindrica f. laciniata

Det. P.W. James   11 19 61

Lichenoides corneum, marginibus eleganter fimbriatis Hist. Musc. p. 218. n. 116.

A3

Lichen pulm. favut. e cineres fuss minimy gris. *Celsii

Lichenoides coriaceum cinereum, peltis atris compressis Hist. Musc. p. 219. n. 117.

Lichenoides rugosum durum pullum, peltis atris verrucosis Hist. Musc. p. 220. n. 118.

119.

Umbilicaria hyperborea

Det. R. Santesson   19 61.

Lichenoides atrum, corii Persici instar exasperatum Hist. Musc. p. 220. n. 119.

E Lapp. a Linnæo.   E Groenlandia.

Lichenoides subtus croceum, peltis appressis Hist. Musc. p. 221. n. 120.

LECTOTYPE

of Umbilicaria torrefacta (Lightf.) Schrad.

left hand specimen of the three numbered 118 (p. 220 Hist. Musc.)

Lichenoides Lichenis facie, peltis acetabulis immersis Hist. Musc. p. 221. n. 121.

In his *Historia Muscorum*, Dillenius attempted to include all the cryptogamic plants known at the time, about 660 species (or rough equivalent) of fungi, lichens, algae, and mosses. The entry for each includes the name, including synonyms (i.e., any different names that earlier authors used for the same organism); a detailed description of distinctive structures; and any medicinal uses. Each species description referenced specimens in his herbarium, although because of the difficulty in preserving them, he maintained paintings rather than specimens for macrofungi. His herbarium contains 901 specimens mounted on 176 sheets. Specimen sheets often contain multiple taxa, each accompanied by both scientific and common names.

Preservation of Plant Knowledge in Non-European Cultures

▲ Micheli's tomb in the Basilica of Santa Croce in Florence, which also contains the tombs of Galileo and Michelangelo.

◀ Moss specimens from the Dillenius herbarium. Recalling the fourth volume of Rauwolf's herbarium, the sheets in this herbarium are typically bordered by thicker, colored paper, purportedly wallpaper remnants.

Although the herbarium originated in Europe, other contemporary world cultures compiled knowledge about plants that surpassed European efforts in volume, longevity, and continuity. The earliest Chinese herbal dates from about 300 BCE, and Chinese scholars produced a new and original work on pharmaceutical botany at least every century between 100 and 1700 CE. Such compendia contained literary as well as popular names and synonyms, detailed descriptions, and accurate illustrations. Each successive publication extended the knowledge base with additional species and additional pharmaceutical uses. The unbroken advance in plant knowledge over time in China led to a consistently high level of accurate description and illustration and a standard method of naming for plants, as well as standardized information about their uses, growth, and distribution.

On the Indian subcontinent, plant knowledge was first recorded in samhitas, ancient Sanskrit texts that document medical practices. The *Sushruta Samhita* and the *Charaka Samhita*, which originated about 500 BCE, describe more than 700 medicinal herbs, including their taste, appearance, digestive effects, efficacy, dosage, and benefits. After the samhitas came the vrikshayurvedas, a type of text on "the science of plants," of which the *Krishi Parashara Vrikshayurveda* from the 1st century CE is the best known. This work formed the basis for the strong development of botanical study throughout the medieval period in India. Following parallel paths in

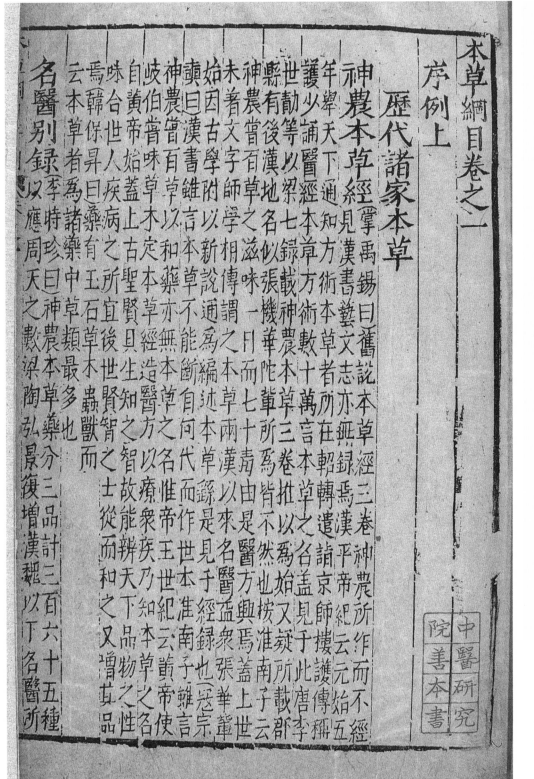

◄ A page from the *Bencao Gangmu* (Compendium of Materia Medica), authored by Li Shizhen in 1578 and held in the rare books collection of the library of the Chinese Academy of Traditional Medicine.

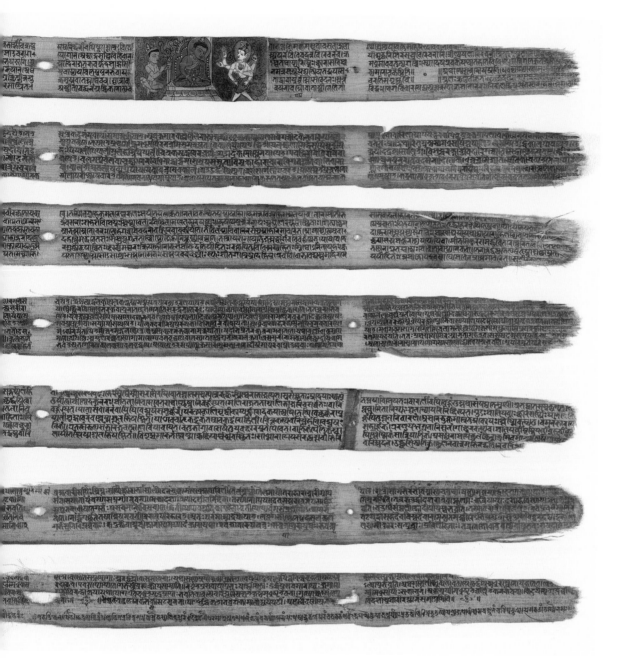

▲ Fragments of *Sushruta Samhita*, a Sanskrit text on medicinal herbs, dating from the 12th or 13th century.

both northern and southern regions of the subcontinent, texts about plants provided information about the natural flora and main agricultural crops, and botanic gardens, which grew plants primarily for medicinal use, supplemented the written record.

The impetus for plant knowledge in the Islamic world had a spiritual as well as practical basis, because descriptions of Paradise in the Qur'an include the names of plants that grow there, and because the plants are invoked frequently in stories of the Creator's power. Abu Hanifa Dinawari (815–896 CE), who lived in what is today western Iran, produced the multivolume *Kitab al-Nabat* (Book of Plants). Only volumes three and five have survived, with part of volume six reconstructed from quoted passages. In what survives, 637 plants are described in alphabetical order, and descriptions include how the plant grows, especially its production of flowers and fruit. Travel was common across the widespread Islamic world, and the botanical literature originally focused on Iran expanded to include species from the Mediterranean region and North Africa. As scholars encountered new species, the literature expanded with new terminology, uses, and observations on plant development. The use and cultivation of plants was documented in the 11th century by Ibn Bassal of Toledo in his book *Dīwān al-Filāha* (Court of Agriculture); and in his 12th-century *Kitab al-Filāha* (Treatise on Agriculture), Ibn al-'Awwam of Seville described over 180 plants (mostly leaf and root vegetables, herbs, spices, and fruit-bearing trees) and how to propagate and care for them.

We know that the Aztecs recorded information about plants, although European invaders destroyed most of the evidence of these works. The Aztec king Montezuma maintained botanical and zoological gardens that were said to be far more advanced than those in Europe. Aztec libraries contained thousands of manuscripts written on paper made from the bark of *Ficus citrifolia* (shortleaf fig). Sadly, the libraries were burned in the European conquest of these territories, destroying all records of pre-colonial scholarship. Most of what we do know about Aztec plant knowledge comes from *Libellus de Medicinalibus Indorum Herbis* (Little Book of the Medicinal Herbs of the Indians), which dates from 1552, some 30 years after the conquest of the Aztecs. It was written by two Aztecs who assimilated into colonial culture, Martinus de la Cruz, a native physician who composed the work in Aztec, and Juannes Badianus, who translated the text into Latin. The Badianus Manuscript, as it is also known, records the indigenous uses, primarily medicinal, of 184 plants, along with illustrations, grouped according to the medical condition to be treated. The illustrations are highly stylized and not accurate in botanical detail but do include simple ecological information about habitats, perhaps as an aid for collecting.

البقلة اليمانية هذا نوع از بقول البيت
كانخورند و مايده دواى ندارد
الملوخيا بقله كه در فعل بناس ميكارند

قاذ ماليس بوصى اوراكنزروشتى
كويند ومعداركنيد بلند شو دنازك شيخ بخ
وبرك اوجوان برك بامل باش برنشعب لكنه
ستى نازك وباريك وبرك اوخوش باش
سنيد سرفى فايل اوراخام وكشته توان خورد

ولمل برله   جن جبير
اوراسباريك محو زرند زهوت جاع بالبركت اورد وتخم او زند
تخم اوراكبرند وتخم او دكشتنهاه شب كه دحل كنند ونكاه دارند كه
وه بجير دستى زركستان لسباروميه درموش كو
براند ولسبارده وبشه وتنرى اورا زياده برشتانى شيخ

بادروج جون لسبارگوزندوكلى جشم اورد
و لشع ورا ليام كنند وها واورا لحركت اورد
ولمل برله نهر بشد لسبان كنند وفرك رسیل اجوان مارد

جنبراب رام جاره غماد كنند   مفيد يا شعدجوان

اوغنى كل سخ نرن ثر شيخ

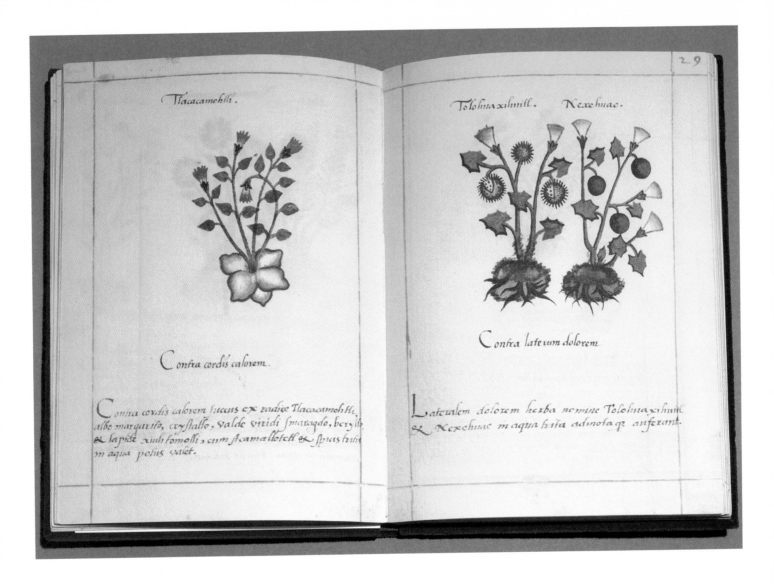

The plant illustrations bear the following handwritten labels: *Tlacacamohtli.* / *Tolohuaxihuitl.* *Nexehuac.* / *Contra cordis calorem.* / *Contra laterum dolorem.*

But of all the world cultures with a long history of plant documentation, Europeans were the only ones to preserve specimens as part of that process. There are several possible reasons for this. The strongly seasonal European climate may have been a factor, because plants change drastically in appearance in winter or disappear entirely, making it impossible to observe them throughout the year. In other parts of the world with a tradition of written scholarship, for example, India, China, Latin America, and the Middle East, the climate is more moderate and plants could be more readily observed for a longer period of the year. Also, areas of the world with a more humid climate might not have been able to keep herbaria free of mold and insect infestation.

▲ Pages from the Badianus manuscript.

◀ PREVIOUS SPREAD Page from *Kitab-i hasha'ish* (Book of Herbs), dating from the late 16th or early 17th century and probably created in the vicinity of the Deccan Plateau.

Perhaps more important was that none of these non-European cultures had a prolonged period of intellectual stagnation like Europe's Middle Ages, which interrupted the tradition of recording natural history information that had been established by classical authors. When science reemerged in the Renaissance, European scientists lacked a knowledge base that had been accruing century by century, forcing them to start afresh to build this knowledge. The lack of precedent and an intellectual climate open to new ideas spawned innovations such as the telescope, microscope—and herbaria.

Adhatoda
Zeylanensium. H.L.
Justicia
adhatoda.
S.

Adhatoda Vasica Nee
CBClarla

Herbaria *and the* Age *of* Botanical Exploration

In the 18th and 19th centuries, European countries, among them Austria, England, France, Portugal, and Spain, invested heavily in sea- and land-based explorations. The goal was to establish colonial outposts where they could exploit local resources for trade and to secure safe and dependable routes to transport goods back home. Many of the products they were keen to obtain were plants, or plant-derived, most notably spices. Since prehistoric times, plant products used for food, drugs, and religious ceremonies had been motivating travel and trade, and as a European middle class emerged, so did the demand for products derived from plants that could not be grown at home. In addition to augmenting their diets, Europeans wanted to enliven their gardens with exotic plants. In the Netherlands, the phenomenon of Tulipomania created such high demand for tulips that the price for a single bulb rose to the equivalent of thousands of dollars before the market collapsed in 1637.

◀ Specimen from the Clifford Herbarium.

◀ TOP TO BOTTOM
Cinnamon is derived from the bark of *Cinnamomum verum*, a member of the laurel family, native to Sri Lanka.

Nutmeg and mace are the seed and aril, respectively, of *Myristica fragrans*, a tree indigenous to Indonesia's Maluku Islands.

The flower buds of *Syzygium aromaticum*, another tree native to the Maluku Islands, are the source of cloves.

▲ Semper Augustus, one of the tulips most prized by Europeans in the 17th century.

Collecting Plants for Herbarium Specimens on Expeditions

Der Botaniker, painted in 1908 by Hermann Kern, depicts a botanist with his metal collecting case, or vasculum, raising a glass to a successful day in the field.

Although the botanical objectives of a government-sponsored exploring expedition usually focused on a few species of commerce, by the beginning of the 18th century the idea emerged that at least some expeditions should include naturalists to document the native vegetation and animals of the foreign lands visited. Such trips resulted in the collection of living plants as well as those preserved as herbarium specimens. Only about 12,000 of the world's estimated 400,000 plant species occur naturally in Europe, so almost every trip brought home plants that no botanist there had ever seen before. As a result of this influx of exotic flora, understanding the breadth of plant diversity became a central focus of botanical study.

English botanist John Woodward's *Brief Instructions for Making Observations in All Parts of the World, as also, for Collecting, Preserving, and Sending Over Natural Things* (1696) laid down the basic principles for plant collection. Those charged with plant collection on an expedition would make sorties from the ship or land-based camp, looking for plants suitable for collection, which would generally be those displaying as many of the features needed for identification as possible, such as leaves, buds, flowers, and fruits. Portions of such plants would be placed in a cloth sack or vasculum, a metal case in the shape of a somewhat flattened cylinder that opened along the long side and had a handle or strap for ease of carrying. Pressing of the specimen would take place as soon as possible after collection.

Methods for preparing plant material for herbarium specimens have not changed significantly from the methods given in Woodward's treatise, although the materials have changed. The procedure begins with the laying out of the plant between sheets of paper in a way that keeps all key parts visible and not overlapping one another. The paper could be any absorbent type—sometimes blank pages were made into quires (sets of four to eight pages bound together) expressly for this purpose. The botanists on Captain Cook's first voyage were said to have used unbound copies of Milton's *Paradise Lost* as pressing papers, but old newspaper is the usual choice

today. The plants in their papers are then stacked in a plant press for flattening and drying. A plant press consists of two solid boards or cross-hatched strips of lathe held together with straps or screws; the press is tightened as the plants dry to make them as flat as possible.

To be of value for scientific research, the specimen must be accompanied by information about where and when it was collected, the collection date, and the name of the collector. The name of the plant is also a critical part of the specimen metadata, although frequently the identity is unknown at the time of collection and is added later. Sometimes a number accompanies the collector's name. The number refers to the place of this collection in the sequence of all those gathered by a given collector. The collection numbers would be recorded in a log known as a field book or field notebook that lists the specimens in order by collection date and locality. Additional information that might accompany a specimen includes the local or indigenous name of the plant and its uses for food, medicine, or art.

Standards for data collection were not as well established as techniques for specimen collection, and early naturalists worked in situations where time and space for

▲ Specimen being prepared for the press out in the field.

▶ An herbarium mounting station showing the tools for the job.

► Botanists on a collecting expedition to Surinam, 1963. Each man is standing behind a plant press, loaded with plant specimens, that has been placed on a wooden frame over a heating element (probably a propane camp stove); a fifth plant press is on the ground. Pictured left to right are Howard Irwin, Ghillean T. Prance, and Noel Holmgren of the New York Botanical Garden, and Thomas Soderstrom of the U.S. National Herbarium, Smithsonian Institution.

processing of collections were limited. Rough seas, illness, hostilities with local people, lack of appreciation of the collecting effort by the expedition leader and non-scientific members—all worked against the orderly documentation of collections. Once dry, the now very brittle specimens had to be stored and protected against water, fire, insects, and mold until the expedition reached home or a suitable waypoint for shipment.

The collection of plants that were intended to be kept alive on shipboard for months or even years required a container in which plants could receive sufficient light and yet be protected from wind, seawater, and drying. Most plants collected on the earliest expeditions did not survive the trip; however, the invention of the Wardian case (developed by Nathaniel B. Ward, 1791–1868) improved the survival rate. The Wardian case was a sort of mobile terrarium in which plants could grow while being transported on shipboard. Robert Fortune, a Scottish botanist employed by the British East India Company, used Wardian cases to smuggle seedlings of tea plants out of China in 1848. Some 30 years later, seeds of the rubber plant, *Hevea brasiliensis*, were purchased from Brazil and sent to the Royal Botanic Gardens at Kew, where they were germinated; the seedlings were then sent in Wardian cases to Sri Lanka to begin the Asian rubber industry.

The best-provisioned expeditions took along artists to capture how a plant looked in living condition, as an accompaniment to collected specimens. Producing accurate and pleasing renderings of plants on shipboard was often extremely challenging, yet some very fine examples exist. One of the finest expedition illustrators was Ferdinand Bauer (1760–1826), an Austrian botanical illustrator and trained botanist who traveled on Matthew Flinders' circumnavigation of Australia (1801–03), as one of six naturalist-illustrators under the direction of botanist Robert Brown. Bauer only sketched the plants during the expedition; he wrote numbers on the sketch that referred to a color table to guide the addition of paint to the sketch when he returned home—an early version of "paint by numbers."

The hundreds of 18th- and 19th-century exploring expeditions yielded plant specimens that are still stored in herbaria today. There are fascinating stories behind all expeditions, but we will focus here on profiling several individuals and expeditions that had a strong influence on later explorers and the development of herbaria.

William Dampier, Privateer and Naturalist

The first explorer charged specifically with the collection of herbarium specimens was William Dampier (1651–1715). Born in southwestern England (East Coker, Somerset County), Dampier started his life at sea shortly after completing his primary

▶ CLOCKWISE, FROM TOP LEFT
Mounting an herbarium specimen at the William and Lynda Steere Herbarium involves gluing the dried specimen to a sheet of 100% cotton rag paper (11 × 17 in.).

Specimen being positioned on the mounting paper. Note that the specimen label has already been glued to the paper, as has a small packet (here weighted down by two metal washers) to hold any fragments that may accompany the specimen.

The specimen is then covered with waxed paper, weighted with metal washers, and allowed to dry overnight.

The next day, woody portions of the plant may be sewn to the sheet to secure the specimen further.

▶ Wardian cases for transporting living plant specimens. The ones shown here are full of cycads from Rockhampton, Queensland, newly arrived at the Missouri Botanical Garden after a long journey via London and New York, c.1920.

Marucuja Baueri.

An engraving of Ferdinand Bauer's drawing of *Passiflora aurantia*,
a passion flower, from John Lindley's *Collectanea Botanica* (1821).

Portrait of William Dampier
by Thomas Murray, c.1697.

education. He signed on to two merchant voyages to New-foundland before a brief stint in the Royal Navy. Afterward he spent some time working on a plantation in Jamaica but hated the near slavery-like conditions of that work, so he joined a crew that harvested logwood timber along Mexico's Caribbean coast. *Haematoxylum campechianum* (Mexican logwood) was widely used as a dye for cloth or paper and is still used for staining tissues in microscopic preparations. Dampier returned to England in 1679, married, then almost immediately left England again on a voyage to the Americas. This time Dampier sailed as a privateer, a sort of government-sanctioned pirate. A privateer had permission from a government (a letter of marque) that allowed him to attack foreign ships during wartime and take the ship and its contents as a prize. Proceeds were divided among the sponsors, shipowners, captains, and crew.

After some time in the Americas raiding ships in the Bay of Campeche and crossing the Isthmus of Panama to raid Spanish settlements in Peru, Dampier embarked on his first circumnavigation. He crossed the Pacific to Guam, then traveled on to the Philippines, Australia (west coast), China, the Maluku Islands, then around the Cape of Good Hope, raiding foreign ships whenever possible. He arrived home, after 12 years' absence, in 1691. What set him apart from the typical person in his line of work was that Dampier kept a detailed written account of the trip, not only of his plundering adventures but also of the lands, people, plants, and animals he encountered. He went to great lengths to maintain his journal, storing the rolls of parchment on which he wrote within hollow bamboo stems, the ends of which he sealed with wax. Dampier was penniless on his return to England, but his journal was intact.

While Dampier was at sea, a new phenomenon had become very popular in England: the coffee house. Seafarers and leaders of the scientific community were among the diverse clientele of these establishments, which provided rare opportunities for socialization across traditional social class lines. At the height of their popularity, more than 2,000 coffee houses operated in London. Important institutions resulted from discussions that took place in such establishments, including the London Stock Exchange and Lloyd's of London. Upon his return Dampier became a regular at Jonathan's in Change Alley, where he met prominent scientists such as Robert Hooke, Hans Sloane (whose collections would later form the nucleus of London's

PLANTS OF BAHAMA ISLANDS

Haematoxylum campechianum L.

Great Exuma: in open coppice just south of
    Sheep Cay.

Small spreading tree 3.5 m tall; flowers
    cream-color.

January 8, 1975

D. S. Correll   44044

Herbarium
FAIRCHILD TROPICAL GARDEN

Haematoxylum campechianum L.

D. McJunkin [LA]   25.i. 198?

An herbarium specimen of *Haematoxylum campechianum* (Mexican logwood).

"JONATHAN'S." *From an Old Sketch.* (*See page* 473.)

▲▲ Mexican logwood shavings produce a rich purple used for dyeing cloth and wool.

▲ Cartoon (c.1881) of Jonathan's Coffee House, located in Change Alley, London.

Natural History Museum), and Robert Southwell, president of the Royal Society, the predominant British scientific organization of the time. No doubt the interest shown by these new acquaintances in his just-completed journey encouraged Dampier to write a book based on his journal, *A New Voyage Round the World*, published in 1697. In it, Dampier gave astute and lucid descriptions of landforms and currents, as well as of the plants and animals encountered during his journey. His descriptions of Australia (New Holland) were among the first such accounts by an English subject.

Dampier's book was very popular with the English public—it went through three printings and was translated into Dutch, French, and German. The English literary magazine *Works of the Learned* recommended the book for the sedentary traveler, praising it for the "variety of its descriptions and surprisingness of its incidents." The success of his first book provided Dampier with the opportunity to dine at the home of Samuel Pepys, noted diarist and holder of a variety of government positions, including Secretary to the Admiralty. Perhaps as a result of that meeting, the Royal Navy invited Dampier to propose a new expedition. He suggested a survey of "terra australis," where he believed deposits of gold would be found. The European concept of this southern continent was based on existing reports of Australia, but many, including Dampier, thought it was probably a much bigger landmass, extending all the way to the South Pole. The Royal Navy accepted his proposal and gave him a commission to explore the eastern coast of Australia. For the first time for such an expedition, the Royal Navy specifically charged Dampier with documenting natural history as well as gathering geographic and navigational data, and he received training in natural history specimen collection from John Woodward. Although enthusiasm was high for the voyage, the ship assigned to it was a little disappointing. The HMS *Roebuck* was a nine-year-old fireship, one that, being near the end of its useful life, was slated to be used as a weapon. In the height of a close naval battle, the ship would be filled with combustible contents, sent into an enemy fleet, and then ignited.

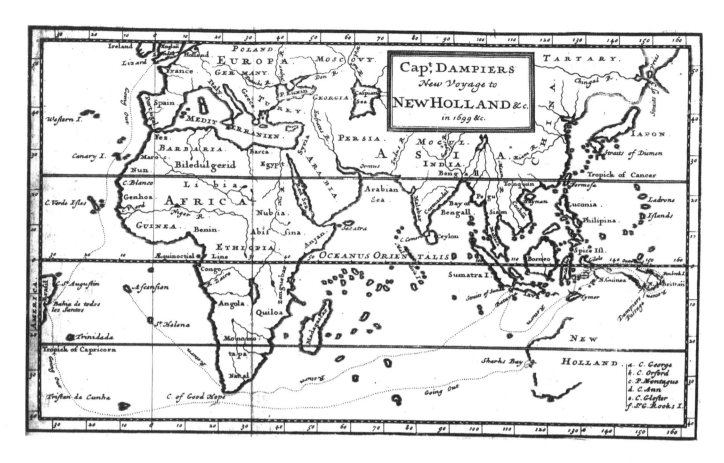

The map shows "Capt. Dampiers New Voyage to New Holland &c. in 1699 &c."

▲ Map showing the route of the HMS *Roebuck* from Dampier's *A Voyage to New Holland* (1703). It sold well but was not the blockbuster his first book was, probably because the emphasis was more on natural history, less on swashbuckling.

◀ Rendering of the HMS *Roebuck* along the New Guinea coast during the Roebuck expedition of 1699. From *The Book of Ships*, published c.1920.

The *Roebuck* departed in January 1699, stopping first at the Canary and Cape Verde Islands, then across the Atlantic to Bahia, Brazil. From the outset Dampier had problems with his crew, especially his second in command, George Fisher, who was openly contemptuous of Dampier's privateering past. Dampier suspected him of plotting a mutiny almost from the outset and so when they arrived in Brazil, he had Fisher taken off the ship in chains and imprisoned in Bahia until another English ship could transport him back home. The original plan was to travel from Brazil around Cape Horn, but by the time he was ready to leave Brazil, it was too late in the season to attempt that passage. So, instead they headed east. They did not stop at the Cape of Good Hope, as most ships heading east did, but instead took a more southern route where Dampier was able to take advantage of the strong winds of the Roaring Forties (the area between the 40th and 50th south parallels) to fill his sails. In August 1699, about four months after leaving Brazil, they reached the western coast of Australia, low on supplies of food and water but without any loss of crew to scurvy or other mishaps.

Dampier's first Australian landfall was Dirk Hartog Island in Shark Bay. This island was named for the first Dutch explorer to set foot there in 1616. The island was visited again just two years before Dampier by another Dutch explorer, Willem de Vlamingh. Unlike these previous visitors, who apparently made no observations or collected any specimens of the distinctive flora there, Dampier immediately began to document the plants and animals he encountered. Of the vegetation he wrote,

> Most of the Trees and Shrubs had at this Time either Blossoms or Berries on them. The Blossoms of the different Sort of Trees were of several Colours, as red, white, yellow, &c., but mostly blue: And these generally smelt very sweet and fragrant, as did some also of the rest. There were also beside some Plants, Herbs, and tall Flowers, some very small Flowers, growing on the Ground, that were sweet and beautiful, and for the most part unlike any I had seen elsewhere.

Indeed, the flora of western Australia is so distinctive that most of the plants Dampier saw were completely unknown to European botanists. Dampier collected 22 flowering plant specimens and one marine alga on Dirk Hartog Island, which he pressed and dried between the pages of books. Although he complained that some specimens were "spoil'd," those that survived were very well preserved and perfectly adequate for scientific study, although rather small. Illustrations were made of some of Dampier's collections, but the artist is not attributed. While not of professional quality, these were accurate enough to give a sense of the growth habit of the plant.

From Dirk Hartog Island, Dampier sailed through Shark Bay, eventually reaching the mainland. At this point the *Roebuck* was nearly out of fresh water, and finding a source for refilling their supply distracted Dampier from his natural history exploration. Unsuccessful in finding drinking water in the arid Shark Bay region, the expedition headed north, following the coast near present-day Dampier Archipelago and Roebuck Bay. On East Lewis Island Dampier discovered his most spectacular flower and an emblematic Australian plant, Sturt's desert pea.

▲▲ Vegetation in Francois Peron National Park, Shark Bay World Heritage Region, across the bay from Dirk Hartog Island.

▲ *Swainsona formosa* (Sturt's desert pea), first collected by Dampier and originally described as *Clianthus dampieri*, a member of the legume family.

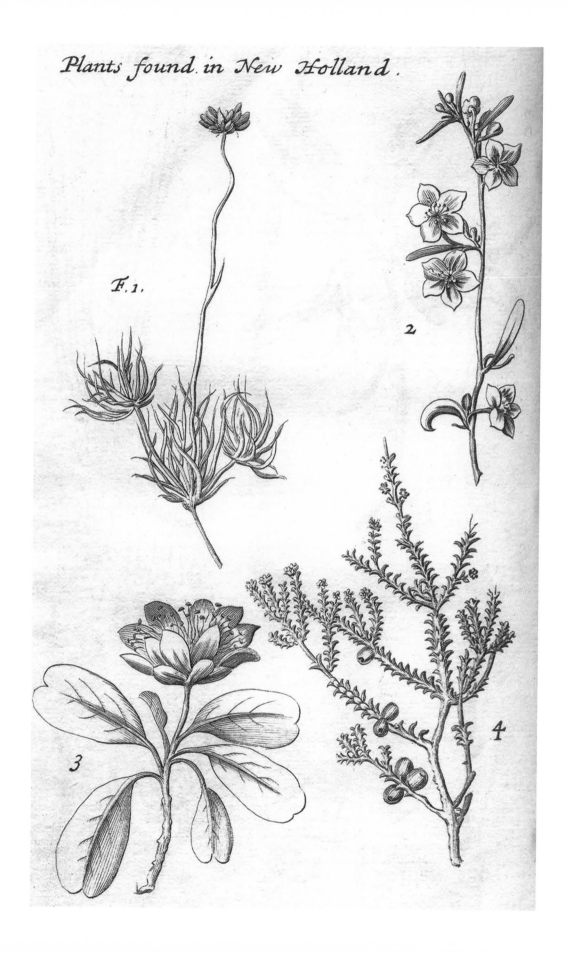

Illustrations of
Australian plants from
Dampier's *A Voyage to
New Holland* (1703).

Swainsona formosa
(G. Don) Thompson

Willdampia formosa (G. Don) A.S. George

Det. *A.S. George*   Date 16 March 2000

Colutea novæ Hollandiæ,
fl. amplis coccineis, um-
-ballatim dispositis, ma-
-cula purpurea notatis.
Dampier voiaga vol. 3.
Tab. 4. fig. 2.   15.
An Coral arboris species?
flos similis fere Agaty inter Decandras

Sherardscript

Dampieri A.Cunn
Clianthus formosus

Clianthus formosus (G. Don) Ford &
Vickery

Determinavit *R. Melville*   1958.

C. Dampier 3 Vol. Tab. 4. f. 2.
N. Holl.
H.S. Willdenow 15

with Ray historia tert supp. p 226
Thes description in Dampier
voyage III p 159-160 in
translation plain

Don spul mann
no. 2. p. 46 V

Dampier formosa

15

▲ Ascension Island,
where the HMS *Roebuck*
sank just offshore.

◀ Type specimen of
Swainsona formosa
collected by Dampier.

Still unable to successfully replenish the ship's water supply, Dampier decided to leave Australia and head north for the island of Timor, where both the Dutch and Portuguese had outposts. There, thanks to the generosity primarily of the Portuguese, his crew was able to resupply and make much-needed repairs to the *Roebuck*. They then sailed northeast to New Guinea and traveled through the Bismarck Archipelago. All along the way Dampier continued to make detailed written descriptions of the biota and people, as well as collections, primarily of plants and shells. As the trip progressed, however, Dampier became increasingly concerned about the state of his ship. The *Roebuck*'s hull was badly damaged by shipworms, and Dampier decided not to continue on to the eastern coast of Australia, his primary destination according to the charge of the expedition, but instead to begin the trip home. He stopped in Batavia (now Jakarta), capital of Indonesia, for more repairs to the *Roebuck*. While there, he made plans to join a convoy of ships heading for Europe, better ensuring the possibility of rescue if the *Roebuck*'s repairs failed. They made it safely to South Africa, then, with the most difficult part of the journey behind them, Dampier and the *Roebuck* sailed on for home alone. However, the *Roebuck* sank off the coast of Ascension Island. They were near enough that the crew made it safely to land, but

Herbaria and the Age of Botanical Exploration | 55

many of Dampier's specimens were lost. The crew lived off the land in the Clarence Bay region on the northwest coast of Ascension Island for several months until they were finally rescued by a passing Royal Navy ship that returned them to England in August 1701.

Dampier did manage to save his papers and some of his specimens from the sinking *Roebuck*. The surviving botanical specimens included a few from Brazil, the plants from Dirk Hartog Island and Rosemary Island (near Port Hedland, Australia), and a moss (*Leucobryum candidum*) from New Guinea. Upon his return to England, Dampier gave the specimens to his mentor John Woodward, who passed some to John Ray and some to Leonard Plukenet, Royal Professor of Botany and gardener to Queen Mary. Ray described nine of the plants collected by Dampier in his *Historia Plantarum* (vol. 3, 1704); Plukenet described another eight in *Amaltheum Botanicum* (1705). After study, the specimens were returned to Woodward, who gave them, with his whole herbarium, to William Sherard at Oxford University, where they are maintained to this day.

Dampier was court-martialed for both the loss of the *Roebuck* and, more consequentially, the mistreatment of George Fisher, who he had imprisoned in Brazil. After his dismissal from the Royal Navy, Dampier returned to life as a privateer, making two additional circumnavigations while pursuing the ships of England's enemies for plunder. There is no evidence that he made natural history collections on these later voyages, nor did he write any more books.

Dampier's recorded observations were very important for explorers who came after him: Joseph Banks, Alexander von Humboldt, Charles Darwin, and Alfred Russel Wallace all studied his writings. Though only a small number of his collections survived, scientists have been able to identify about 45 species of plants, 16 birds, and 30 marine animals from his descriptions and illustrations alone. Perhaps even more important were his contributions concerning currents, winds, and tides across the world's oceans—later navigators such as James Cook depended on them. His life events inspired fictional adventure stories, including Daniel Defoe's *Robinson Crusoe* and *Gulliver's Travels*, in which Jonathan Swift mentions Dampier explicitly, comparing his skills as a navigator to his fictional protagonist Lemuel Gulliver. Additionally, according to the Oxford English Dictionary his writings contain the first known written usages of more than 1,000 words, including barbecue, avocado, and chopsticks. His extraordinary achievements are commemorated in a number of place names, including Dampier Strait (Indonesia), Mt. Dampier (New Zealand), and Dampier Peninsula (Australia).

After their initial study, Dampier's collections were largely forgotten. The names given by Ray and Plukenet were superseded by ones proposed by Linnaeus

Divers photographing the giant clam shell found among HMS *Roebuck* artifacts.

or later authors. Sturt's desert pea, which was ignored or never seen by Ray or Plukenet, wasn't named until 1832, based on collections made in 1818. Other plants have been named for Dampier, most fittingly *Dampiera*, a genus of blue-flowered shrubs in Western Australia. The beleaguered *Roebuck* is commemorated in *Roebuckia cheilocarpa*, an Australian daisy. On the 300th anniversary of Dampier's visit to Australia in August 1999, Oxford University loaned his specimens for an exhibition at the Western Australian Museum in Perth. As part of the commemoration, a group of Australian and English botanists visited Dampier Landing on Dirk Hartog Island and re-collected most of the plants in Dampier's collection. Collecting plants from the same location 300 years apart provides a valuable opportunity not only to compare which species occurred then versus now but also shows how plants of the same species may have changed over time. Three sets of specimens were made, one for Oxford University, one for the Western Australian Herbarium, and one for a new museum at Denham, a small town on Peron Peninsula in Shark Bay—a fitting tribute to Dampier's remarkable accomplishments and an important piece of Australian history.

In 2000, an effort led by the Western Australian Maritime Museum located the remains of the *Roebuck* in Clarence Bay on Ascension Island. They were able to find a ship's bell and various bits of pottery, all consistent with the age and fittings of the *Roebuck*. Most artifacts were left in place, but those that were loose and in danger of being swept away by currents were salvaged, and replicas of these are on display in the Western Australian Maritime Museum in Freemantle. In among the *Roebuck*'s remains on the sea floor the divers found a giant clam shell belonging to a species of the genus *Tridacna*, which is native to the tropical and subtropical Indo-Pacific region, thousands of miles away from Ascension Island. Dampier wrote about finding giant clams in *A Voyage to New Holland*, and so it is possible that this shell is one of Dampier's collections that did not make it home!

Carl Linnaeus, Architect of Botanical Infrastructure

Although he did not travel widely, Carl Linnaeus was a critical figure during the age of botanical exploration. His breadth of vision, compulsion to organize and teach, and generous measure of self-confidence allowed him to construct an intellectual and practical framework for the study of plant diversity. The infrastructure he put in place facilitated the incorporation of new knowledge about plants as it was discovered and created enduring standards for documenting that knowledge. Linnaeus is a towering figure in the study of plants, animals, and minerals; his life and work are very well documented, and so we will focus here on his contributions as they directly relate to the preservation of plant material in herbaria.

Linnaeus was born in Småland, Sweden, in 1707. His interest in botany began in his youth, when he often ignored his other studies in order to explore the local countryside for plants. While a student of medicine and botany at Uppsala University, Linnaeus made his only major collecting expedition, to Lapland, the northernmost region of Scandinavia. Linnaeus' goal in this journey was to collect plants, animals, and minerals of potential commercial importance and to observe the habits of the Sami, the native people of the region. He left on his six-month trip in 1732, traveling on foot or by horseback, carrying with him his journal, manuscripts, and collecting implements.

He summarized the botanical findings from his Lapland trip in *Flora Lapponica*, a publication that included 534 species, 100 of which were not known previously. This book became the model for the way a botanist shares knowledge of the plant diversity of a given region, and such a publication is today still called a flora. In this work he also debuted a new classification system for plants. This classification system was based entirely on plant sexual organs—the number of stamens (male) and their position relative to the pistil (female), whether flowers have one or both types of sexual organs, and so forth.

Linnaeus' classification system caused a ruckus in academic circles. It was a significant departure from the two predominant classification systems at this time,

▲ Replica of a portrait of Linnaeus in his Lapland journey attire by Hendrik Hollander, 1823. In his hand is *Linnaea borealis*; the genus is named for him.

▶ Georg Dionysius Ehret's illustration of Linnaeus' sexual system of plant classification. This diagram appeared in the first edition of *Systema Naturae* (1735); the original watercolor is held in London's Natural History Museum.

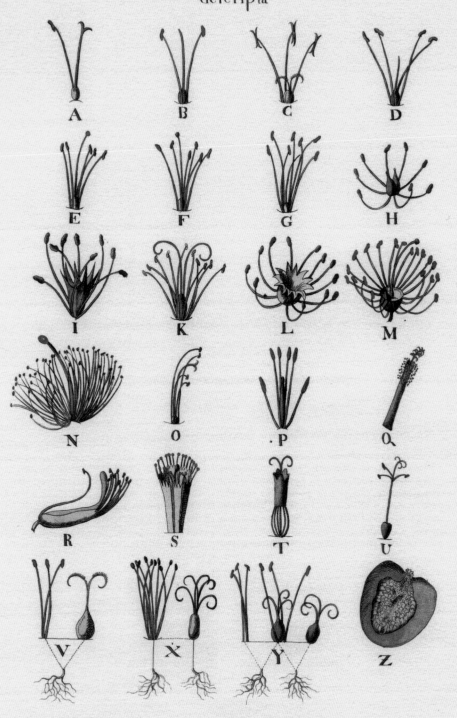

Clariss: **LINNÆI. M.D.**

METHODUS plantarum **SEXUALIS**

in *SISTEMATE NATURÆ*

descripta

Lugd. bat: *1736*

G.D. EHRET. Palat-heidelb:

fecit & edidit

those of Ray and de Tournefort, which used a wider range of plant features to group plants, including stature (e.g., herb, shrub, or tree) and leaf type as well as flower characteristics. Some thought the emphasis that Linnaeus placed on sexual organs was indecent, as were his analogies between plant and human reproduction, such as describing the flower as a bridal chamber where copulation with one or more partners might take place. Botanist Johann Siegesbeck called it "loathesome harlotry." Linnaeus later gave the name *Sigesbeckia* to a genus of rather unattractive weedy plants.

QVADRVXIO MARGARITARIA

▲ One of the small slips of paper that Linnaeus used to record information about organisms. On this card, he summarizes, in Latin, the key features of *Margaritaria*, a small genus of tropical trees.

▶ Title page of *Systema Naturae*, first edition, 1735.

Undeterred by criticism, Linnaeus continued to promote his classification system for the rest of his life. In 1735 he published the first edition of his *Systema Naturae*, a classification of all living things, not just plants. A 12-page paper in its first printing, this work by its 10th edition in 1758 would include 7,700 plant species and 4,400 animal species. Through this work Linnaeus provided a continually updated inventory of life that incorporated the novelties that explorers brought back from their voyages. To keep track of the new organisms he encountered between editions of his *Systema*, Linnaeus recorded information about each on small rectangles of paper that could be stacked and rearranged, thereby inventing the index card, the simple device that people everywhere used to organize information before the age of electronic databases.

Also in 1735 Linnaeus moved to the Netherlands, where he quickly completed a doctoral degree in medicine before returning to the study of natural history. Afterward, George Clifford III (1685–1760), a director of the Dutch East India Company and owner of an extensive Dutch estate and garden, hired him as his physician and herbarium curator. Clifford's herbarium contained about 3,500 specimens, derived from his own garden as well as plants from Dutch outposts around the world. Breaking with tradition, this herbarium was maintained on separate sheets rather than bound in book-like volumes. Another distinctive feature of Clifford's herbarium is that many of the specimens have engraved drawings of vases or urns glued at their base, positioned to make it look as though the plant were a bouquet! Linnaeus stayed with Clifford for several years, documenting the botanical holdings in Clifford's herbarium and botanic garden in his *Hortus Cliffortianus* (1737). Joseph Banks purchased the herbarium after Clifford's death, and it became part of the holdings of the Natural History Museum in London, where it remains.

CAROLI LINNÆI, *SVECI*,

DOCTORIS MEDICINÆ,

SYSTEMA NATURÆ,

SIVE

REGNA TRIA NATURÆ

SYSTEMATICE PROPOSITA

PER

CLASSES, ORDINES,

GENERA, & SPECIES.

O JEHOVA! Quam ampla sunt opera Tua !
Quam ea omnia sapienter fecisti !
Quam plena est terra possessione tua !

Psalm. CIV. 24.

LUGDUNI BATAVORUM,
Apud THEODORUM HAAK, MDCCXXXV.

EX TYPOGRAPHIA
JOANNIS WILHELMI DE GROOT.

Linnaeus returned to Sweden in 1738 and never left again. He married and settled in Stockholm to work as a physician. He stayed in Stockholm long enough to found the Royal Swedish Academy of Sciences and serve as its first president, but left three years later to become professor of medicine at Uppsala University. He eventually traded the medical teaching duties for supervision of the university's botanic garden and teaching botany and natural history. To help him expand the *Systema Naturae* he arranged for some of his most devoted students, who he referred to as his apostles, to carry out far-reaching botanical explorations. Pehr Kalm, from Finland, spent two and a half years exploring the flora and fauna of Canada and the U.S. mid-Atlantic region. Daniel Solander traveled with Joseph Banks on the expedition led by James Cook on the *Endeavour* (1768–71) and later worked at the Natural History Museum in London. Carl Thunberg, who spent nine years exploring South Africa and Japan, provided Linnaeus with the first collections from those areas. Other Linnaean students had less success—six of them died of accident or disease while on plant-collecting expeditions.

Linnaeus' most important work for botany was *Species Plantarum*, first published in 1753, in two volumes. It went through five editions, the last two published after Linnaeus' death. This was the first work to consistently name organisms by a two-word combination: the genus and species. Previously, scientists referred to an organism by a short descriptive phrase in Latin, the first word of which denoted what group the organisms belonged to, followed by key features that would distinguish it from others in its group. Linnaeus kept the first word of the phrase, the genus, but shortened the distinguishing phrase to just a single word—the species epithet. He placed the genus and species names at basal ranks in a classification hierarchy that begins with kingdom. Linnaeus' binomial system of nomenclature became standard for both plants and animals because it made each name distinctive and easier to remember.

Species Plantarum included information for about 6,000 species of plants. This information included not just the genus and species names but also a short description of its distinguishing features and a list of other names (synonyms) that previous botanists had used for it. In 1905, because of its consistency and comprehensiveness, *Species Plantarum* was chosen by attendees at the second International Botanical Congress in Vienna as the starting point for botanical nomenclature, which meant that any names for plants given by authors before 1753 were ignored in favor of the name for that plant used by Linnaeus. If a plant was not included in *Species Plantarum*, a new name must be published for it.

Linnaeus strongly promoted the herbarium as the central resource for the study of plant diversity. In his *Philosophia Botanica* (1751), a work that is often described

Specimen from the Clifford Herbarium, with characteristic paper urn glued to plant base. The color target, seen here and elsewhere, is a standardized color reference used to capture optimally exposed and accurately color-balanced images under any illumination.

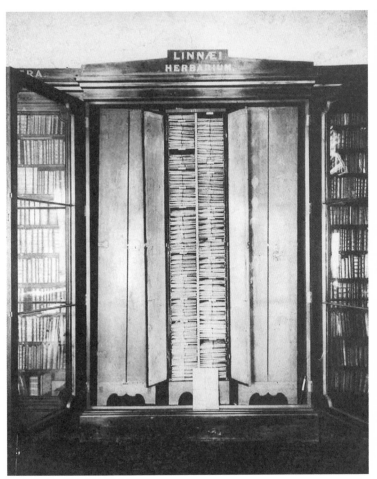

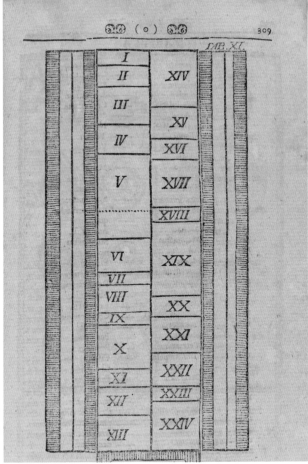

as the first textbook of botany, he wrote: "A herbarium is better than any illustration; every botanist should make one." In this work, he described how plants should be collected, pressed, dried, and glued to paper, and how to store them in an herbarium. Following the procedure he learned from George Clifford, Linnaeus recommended mounting a single specimen per sheet of paper, and to keep these sheets separate. This way, specimens of related species could be kept together even as new collections are obtained, by intercalating the new with the old. He included a diagram of his bespoke herbarium cabinet—a narrow wooden case with 24 shelves, noting which plant group should be placed on which shelf. The major shortcoming of Linnaeus' rules for herbarium storage was his grouping of all cryptogamia into one shelf at the bottom of the herbarium case. Linnaeus was not much interested in algae, bryophytes, and fungi and clearly didn't study them carefully—indeed many of the names he included for these organisms came from Micheli or Dillenius.

Of himself, Linnaeus wrote in one of several autobiographies: "No one has been a greater botanist or zoologist. No one has written more books, more correctly,

▲ Diagram from *Philosophia Botanica* showing the recommended organization of specimens within an herbarium case.

▼ The three herbarium cases of Linnaeus, inside a larger cabinet at the Linnean Society in London. The middle case is open to reveal the specimens in bundles on the shelves. Two of these cases were later returned to Sweden.

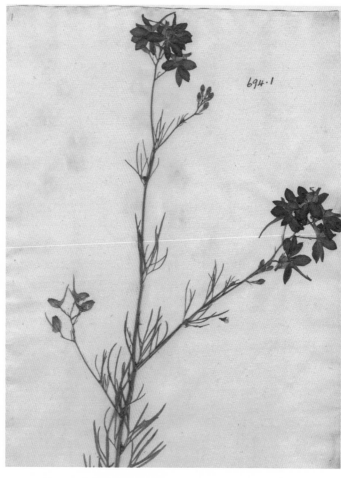

Superbly prepared specimens of *Linnaea borealis*, Linnaeus' favorite flower, and *Delphinium grandiflorum* from the Linnean Herbarium. The numbers (792-1 and 694-1, respectively) refer to the position of the species in his classification.

more methodically, from personal experience. No one has more completely changed a whole science and started a new epoch." While characteristically immodest, this self-assessment of his accomplishments is mostly accurate. His classification system for plants, which he promoted throughout his career, did not endure—his successors preferred a system more closely aligned with that of John Ray. However, by setting the standard for how we name organisms, how we document and share information about their diversity, and how we preserve organisms as specimens for future study, he, more than any other single person, is responsible for the organized archive of life on earth over the past several centuries that we have today. For plants, that archive is the herbarium, and because of Linnaeus, it is permanently relevant to the study of plant diversity.

At the time of his death in 1778, Linnaeus' herbarium consisted of about 14,000 specimens. It was purchased by James Edward Smith, a wealthy amateur botanist, founder of the Linnean Society in 1788 and author of *The English Flora* (1824–28). The herbarium reached England in 1784, along with the three wooden cabinets in which it was stored; also included was Linnaeus' library, as well as his collections of minerals, dried fishes, reptiles, and insects. Two of the wooden cabinets were later returned to Sweden and are stored in a museum adjacent to the old botanic garden of Uppsala.

Today Linnaeus' collections are kept in a carefully climate-controlled underground facility at the home of the Linnean Society. They are available for consultation by scientists and have been digitized and made available online. Linnaean specimens remain relevant for biodiversity science, including type specimens for the 5,900 plant and 4,378 animal specimens that Linnaeus described in *Species Plantarum* and *Systema Naturae*. By rules established for zoology in 1843 and for botany in 1867, every species must be represented by a physical specimen that is dead and permanently preserved in a museum or herbarium collection. Consistent with the scientific method, which dictates that scientific conclusions be verifiable by others, the selection and preservation of type specimens ensures that future scientists can restudy them to either confirm the assessment of the person who originally named and described the species, or come to a different conclusion.

◄ The Linnean collections room, in an underground vault in the library building.

▶ The Linnean library reading room at the Linnean Society, London.

The Bougainville Expedition, or Government-Sponsored Exploration with a Twist

As the 18th century progressed, so did the tempo of overseas exploration by European countries, building on the continually improved navigational information provided by previous travelers, and the ever-increasing demands for goods that could only be obtained from colonial outposts. The Seven Years' War (1756–63) was a global conflict that grew out of the competing colonial interests of European powers. Hostilities spread from North America (where the conflict is usually called the French and Indian War) to the Philippines. Eventually England and its allies (Prussia, now part of Germany chief among them) prevailed over the French and Spanish. For France the loss of colonial territory was particularly acute and spawned a renewed interest in exploring unknown parts of the world in hopes of acquiring new territory.

Despite its international problems, France at the time was enjoying the Age of Enlightenment, a progressive intellectual climate in which Voltaire and other philosophers advanced such ideas as religious tolerance, separation of church and state, and constitutional government. Science was held in high regard by Enlightenment thinkers, who generally agreed that observing, describing, and measuring natural phenomena would contribute to the material and intellectual advancement of humankind. French scientists such as Georges-Louis Leclerc and members of the Jussieu family were contemporaries of Linnaeus and like him were engaged in recording and organizing information about the earth's biota.

From France's thirst to regain colonial territory and a desire to advance knowledge for the good of all people grew the idea for an expedition sponsored by Louis XV. The purpose was to explore the lands between the western coast of North America and the East Indies in the hopes of finding and claiming "terra australis" (elusive since Dampier's day) and any other lands that had not already been claimed by another European nation. The expedition, charged with documenting the plants, animals, and people encountered, was led by Louis-Antoine de Bougainville (1729–1811), an aristocrat with many years of military and diplomatic service and scientific credentials as well. As a young man he published a work on integral calculus that earned him membership in the Royal Society, a governmental body that advised the king on scientific matters.

Philibert Commerson was appointed naturalist on the voyage. His charge was to bring back samples and drawings of everything he considered worthy of attention. Born in 1727 near Lyon, son of a lawyer and landowner, Commerson trained as a physician, but his passion was plants. With encouragement from a teacher while he was a young teen, Commerson began to collect and preserve specimens, and by the

PHILIBERT COMMERSON
Médecin Botaniste et Naturaliste du Roi,
Chevalier de l'Ordre de St Michel,
Membre de l'académie des Sciences.

Imp.Lemercier et Cⁱᵉ.Paris

Philibert Commerson (1727–1773) by Pierre Pagnier, part of the portrait collection in the library of the Natural History Museum, Paris.

age of 15 had compiled a multi-volume herbarium that included detailed observations as well as specimens. He studied medicine at the University of Montpellier, which had a long tradition of botanical study, beginning with Rondelet. There Commerson managed to complete his medical training while spending most of his time in botanical studies, collecting extensively in the Pyrenees and Alps and becoming an acquaintance of Voltaire, who was living in Geneva, Switzerland, at the time. He also gained the attention of Linnaeus, who asked him to investigate the marine life of the Mediterranean for *Systema Naturae* and provided him with an official request from the Swedish government for this work. Eventually Commerson settled into a career as a physician but continued to spend a lot of his time on plants, building a botanic garden in his home in Toulon-sur-Arroux in the Loire Valley. He was recommended for the Bougainville expedition by Linnaeus and the queen of Sweden, as well as by influential French scientists.

Commerson's terms of service with the Bougainville expedition stipulated that he could bring an assistant of his choosing. His unusual and illegal choice was his mistress, Jeanne Baret, born in 1740 to a French peasant family living near Commerson's home. French peasants in the mid-18th century did not benefit much from the egalitarian thinking of the Enlightenment. Most were illiterate and were bound to live and work on land owned by others. They rarely traveled more than 20 miles from their birthplace, and life expectancy was less than 30 years. Baret had been able to lift herself from some of the restrictions of peasant life, however, by becoming an herb woman, a specialist in knowledge of local plants and their medicinal uses. Herb women, who passed their practical botanical knowledge through the generations orally, often had more financial and social opportunities than others. They had an independent source of income from selling medicinal plants to practitioners and apothecaries and were respected for their special expertise.

It may have been Baret's skills as an herb woman that first brought her to Commerson's attention, as a complement to his own academic training in botany and medicine. After a leg injury he sustained doing fieldwork, Commerson was treated successfully with an herbal treatment at a Carthusian monastery in the French Alps, the same one that originated and still produces the herb-infused liqueur Chartreuse (which is also the name of the monastery). Baret became Commerson's lover, living openly with him after the death of his wife, affronting the sensibilities of

Commerson's family and local aristocrats. Baret and Commerson eventually moved to Paris, where living arrangements such as theirs were more commonplace—Commerson's friend, the philosopher Jean-Jacques Rousseau, lived there openly with his working-class mistress. In Paris, Commerson pursued his research at the botanic garden and entered into the intellectual scientific life of the city. From their apartment near the garden, Baret helped him with his research. She also gave birth to a baby boy there but did not keep the child. She gave him to a Paris foundling hospital, where he was later adopted, but he died at a young age.

When it came time for Commerson to depart on the Bougainville voyage, he and Baret hatched a plan to have her join the expedition disguised as a man. This was the only possible way for her to accompany him: a royal ordinance of 1689 forbade women from sailing on French naval ships, and sailors had strong superstitions that a woman on shipboard meant bad luck for the voyage. So, because they could not bear to be apart or because they wanted to continue their scientific collaboration, or both, they found a way to deceive the expedition leadership. Shortly before departure, Commerson came aboard and settled in to his quarters. Baret appeared a little later, impersonating a young sailor looking for work on the expedition. Commerson

A View of Paris from the Pont Neuf, Nicolas-Jean-Baptiste Raguenet, 1763.

Dall'Acqua inc.

MAD.^LLA BARE.

Jeanne Baret, dressed as a man. From *Navigazioni di Cook pel grande oceano e intorno al globo* (1816–17) by James Cook. The plate appears between pages 204 and 205.

pretended to interview her for the job as his assistant, and, after satisfying the captain that "he" was a good fit for the job, hired Baret as his assistant. "Jean," as she was called, came aboard and moved into Commerson's quarters.

The Bougainville expedition consisted of two ships, the *Boudeuse* and the *Étoile*. The *Boudeuse*, upon which Bougainville sailed, was the lead ship. Commerson and Baret sailed on the *Étoile*. This was the less prestigious vessel, a smaller and older one, but perhaps Commerson and Baret thought it would help their deception to keep a distance from the captain. The two ships started out separately, meeting in Rio de Janeiro in June 1767 to begin their joint exploration. They stayed in Rio for about a month, and then traveled south to the La Plata River, where they explored both the Uruguayan and Argentinian shores. The voyage then headed south to Cape Horn, where they made landfall at several sites on the treacherous passage through the Strait of Magellan. From there, the expedition made the long voyage across the Pacific, landing in Tahiti in April 1768, malnourished and suffering from scurvy. They thought they were the first Europeans to land in Tahiti, not knowing that a British expedition led by Samuel Wallis had landed there less than a year earlier, claiming the island for George III. Bougainville called the island Nouvelle Cythère, after the reputed birthplace of Aphrodite in Greek mythology.

The stop in Tahiti was restorative for at least most members of the expedition. The Tahitians were generous hosts, offering an abundance of food, beautiful natural areas to enjoy, and sex—many of the accounts of the voyage described the beauty and friendliness of the Tahitian women in especially glowing terms. After Tahiti, the expedition sailed westward through Samoa and New Hebrides (Vanuatu), thence

to the Solomon Islands and New Guinea. Due to resistance by the residents, the expedition did not make landfall at any of these places, but Bougainville did name one of the largest islands of the group for himself: Bougainville Island, now part of Papua New Guinea. This mountainous island, which is just slightly smaller than Jamaica, was later the site of a prolonged conflict between allied and Japanese forces during World War II. The expedition then pushed on to Java, where they stayed for approximately a month before beginning the trip home.

Baret and Commerson carried out exploration and collection of plant and animal specimens at every opportunity. Commerson was frequently unwell during the voyage, either from seasickness or from his old leg injury, which became aggravated during the voyage and made it difficult for him to walk. He was also frequently dissatisfied. He complained that space provided him was not adequate for the study and preparation of collections, and he had poor relations with some of the other officers, especially with the ship's surgeon, François Vivez, who Commerson suspected was trying to poison him. Baret took on most of the burdens of fieldwork, hiking long distances in pursuit of plant specimens while carrying a musket, a supply of food, and a field press and paper for the press. Her physical strength and tenacity earned her the respect of her fellow travelers, and Commerson referred to her as his "beast of burden."

Baret kept up her disguise as a man until near the end of the expedition. She dressed like a sailor, binding her breasts with strips of linen, a practice that resulted in

A fanciful illustration of the Bougainville expedition: "Monsieur Bougainville Hoisting the French Colours on a small Rock near Cape Forward, in the Streights of Magellan" from David Henry's *An Historical Account of All the Voyages Round the World, Performed by English Navigators* (1773).

View of Bougainville Island from offshore. Although Bougainville attached his name to this island, previously unknown to Europeans, his expedition did not make landfall here, fearing a hostile response from the inhabitants.

serious skin infections from prolonged contact with the rough, sweat-soaked cloth. She bathed, dressed, and urinated away from the other members of the crew, a practice she explained by claiming that she had been made a eunuch during imprisonment in an Ottoman prison in previous military service and was ashamed of her mutilation. Accounts differ as to when her true identity was revealed. Some members of the expedition (including Commerson's nemesis, Vivez) later claimed that they knew Baret was a woman from the start. Bougainville may also have had suspicions early on, because either while at Rio or shortly after leaving, he ordered Commerson confined to quarters for a month, an order that was never fully explained. Bougainville would certainly have wanted to conceal the whole affair as much as possible. The revelation of a woman on the crew would have called his leadership into question and would have detracted from the success of the expedition.

Bougainville's version of the exposure of Baret's true identity, as recorded in his diary, was that it happened in Tahiti. He claimed the Tahitians immediately recognized that Jeanne Baret was a woman, exclaiming about it loudly, and thereby her gender became known by the whole expedition. When Bougainville called upon Baret for explanation, she tearfully gave a story that absolved Commerson of any blame, claiming she had deceived him as well as all other members of the crew. Bougainville didn't believe the story but admired her determination and chose not to punish her. He felt there was little need to make an example of her because, given the difficulty of her life on the expedition, "her example will hardly be contagious." Reports given by Vivez and other expedition members place the revelation of Baret's identity on New Ireland (now Niu Ailan, part of Papua New Guinea) with more violent consequences. In this version of the story, which Baret's biographer Glynis Ridley thinks is closer to the truth, Baret was raped by one or more of her crewmates after her gender was revealed.

Bougainville made the decision that Commerson and Baret would leave the expedition in Île de France (now Mauritius). One story is that upon arrival there, the administrator of the colony made an official request for Commerson to help him carry out botanical explorations. It may also be that the idea to leave Commerson

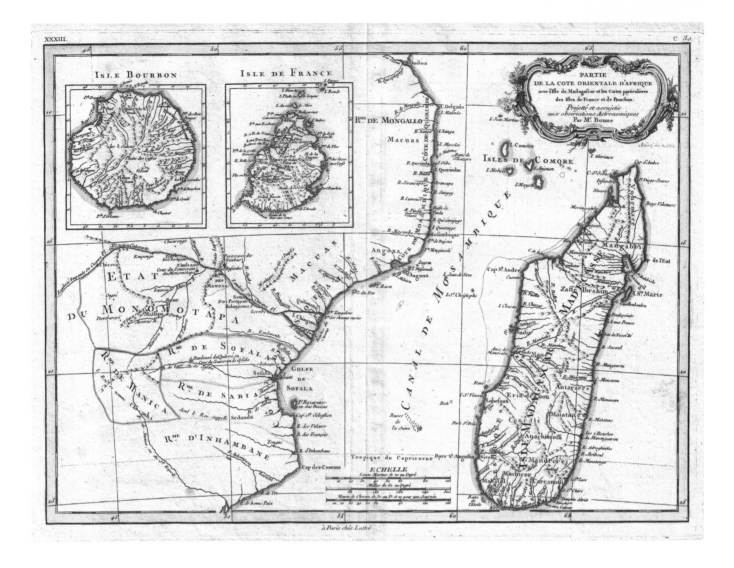

and Baret on Mauritius came from Bougainville as a convenient way to keep the triumphant return of his expedition untarnished by the revelation of Jeanne Baret's true identity. Since Baret was noticeably pregnant by the time they arrived in Mauritius, such a discovery was a foregone conclusion. In any case, Commerson and Baret and all their collections stayed behind when Bougainville and the two ships left for Europe in December 1768. Bougainville's efforts to protect his reputation and that of the expedition appear to have been successful, because the voyage arrived home to great fanfare and was considered a great success by the French government.

Commerson's official duties after leaving the expedition were to report on the natural resources of Mauritius, Réunion, and Madagascar, paying special attention to plants of medicinal value. Baret—after giving birth and relinquishing her second child for adoption, this time to a French couple in Mauritius—continued to

Map of southeastern Africa, including "Isle de France" (now Mauritius), created in 1770 by Rigobert Bonne (1727–1794).

74 | *Herbaria and the Age of Botanical Exploration*

Bust of Pierre Poivre (1719–1786) in the Sir Seewoosagur Ramgoolam Botanical Garden, Pamplemousses, near Port Louis, Mauritius.

conduct fieldwork with Commerson, probably also continuing to do the most strenuous work, since Commerson's health did not improve. They were supported and sometimes accompanied in their collecting trips by Pierre Poivre.

Born in France, Poivre spent most of his life in Asia, first as missionary, then as a member of the French East India Company, eventually assuming oversight of the French colonies in the Indian Ocean (Mauritius, Réunion, and the Seychelles). Trained as a botanist and horticulturalist, Poivre established the botanic garden in Mauritius, now the Sir Seewoosagur Ramgoolam Botanical Garden, oldest in the southern hemisphere. Poivre spent years traveling throughout Southeast Asia and the Pacific looking for seedlings of nutmeg and cloves to take back to Mauritius for cultivation. At the time the only source of these highly valued spices were the Maluku Islands of Indonesia, and the Dutch tightly controlled access to the region. Whether or not it was his idea to have Baret and Commerson join him in Mauritius, he may well have seen their arrival as the opportunity to continue that work.

Poivre's memoir, *Voyages d'un philosophe, ou observations sur les moeurs et les arts des peuples de l'Afrique, de l'Asie et de l'Amérique* (Travels of a Philosopher, or Observations on the Manners and Arts of the Peoples of Africa, Asia and America), published in 1769, was read widely, including by Thomas Jefferson. Unlike most European travelers at the time, Poivre did not consider a priori that European culture was superior to that of others, and he believed that Europe could benefit from the agricultural innovations he encountered. His description of rice cultivation in Vietnam apparently influenced Jefferson to promote this crop in the U.S. Southeast. Another claim to fame for Pierre Poivre is that he is thought to be the protagonist of the "Peter Piper" tongue twister:

Peter Piper picked a peck of pickled peppers.
A peck of pickled peppers Peter Piper picked.
If Peter Piper picked a peck of pickled peppers,
Where's the peck of pickled peppers Peter Piper picked?

His surname translates to "pepper" in English, and peppers are in the genus *Piper*. The earliest printed version of this tongue twister was in *Peter Piper's Practical Principles of Plain and Perfect Pronunciation*, published by John Harris in

London in 1813, which includes a tongue twister for each letter of the alphabet, all basically of the same structure as the Peter Piper verse. However, the rhyme was apparently known at least a generation earlier, and possibly reflects Poivre's efforts to grow pepper and other spices in the French oceanic islands. Green peppercorns were often preserved in liquid, or "pickled" for import.

Commerson's health continued to decline throughout the years in Mauritius, and he died there at the age of 45 in 1773, only a few weeks after his last field trip. Baret continued to assist him until he died, and afterward, being without an income or a place to live, opened a tavern in Port Louis, the capital of Mauritius. The following year she married a former French army sergeant, Jean Dubernat, and returned to France with him, becoming the first known woman to circumnavigate the earth. Her homecoming was not officially noted, but in 1785 she was granted a pension that came with a letter signed by the minister of the French Navy. The document granting her this pension recognizes her accomplishments as follows:

> Jeanne Barré, by means of a disguise, circumnavigated the globe on one of the vessels commanded by Mr de Bougainville. She devoted herself in particular to assisting Mr de Commerson, doctor and botanist, and shared with great courage the labours and dangers of this savant. Her behaviour was exemplary and Mr de Bougainville refers to it with all due credit. . . . His Lordship has been gracious enough to grant to this extraordinary woman a pension of two hundred livres a year to be drawn from the fund for invalid servicemen and this pension shall be payable from 1 January 1785.

With the pension, plus a bequest from Commerson's will, Baret and her husband lived comfortably in Dubernat's home village of Saint-Aulaye (now part of Saint-Antoine-de-Breuilh), near Bordeaux, until her death in 1807.

After Baret and Commerson arrived in Mauritius, Poivre instructed his assistant Paul Philippe Sanguin de Jossigny to sort their previously made collections and notes and make illustrations of selected specimens. After Commerson's death, Jossigny accompanied the specimens back to France, where he delivered them to the Jardin du Roi in 1773. The shipment contained more than 30 cases that were described at the time as including 30,000 specimens representing 5,000 species, of which 3,000 were considered new to science; the shipment also contained 5,000 seed collections. This was by far the largest collection of natural history specimens ever made on an expedition. In 1789, Antoine de Jussieu, botanist at Jardin du Roi, described 37 new plants based on Baret and Commerson collections, though in the publication only Commerson's name was cited as collector. The most famous of the plants described

▲ The village of Saint-Aulaye, on the banks of the Dordogne River, where Jeanne Baret lived after her return from Mauritius. She is buried in the cemetery of the 15th-century church of St. Aulaye.

◀ Grave of Jeanne Baret. The plaque, decorated with paintings of *Solanum baretiae*, recognizes her for being the first woman to circumnavigate the globe.

in this publication is *Bougainvillea spectabilis*, an extremely popular ornamental vine that is now grown worldwide in warm climates. For technical reasons the name has a somewhat complicated history, but the original collection (the type specimen) is the one collected by Baret and Commerson near Rio de Janeiro in Brazil.

A review of known Baret and Commerson collections includes specimens from all the major ports of call of the Bougainville expedition, with the notable exception of Tahiti. Collections made in Madagascar and the other French oceanic islands appear to outnumber those from earlier collection sites, indicating that the Mauritius-based years were more productive botanically than the years aboard the *Étoile*. The database of specimens deposited at the National Herbarium in Paris records about 4,000 specimens of plants collected by Commerson; the specimens are cryptically labeled, but when a collector name is present, it is Commerson's, never Baret's. Commerson specimens are now widely distributed to other European institutions and even to

North and South America, where specimens can be found in the herbaria of the New York Botanical Garden and the University of Buenos Aires.

More than 70 species of plants, birds, and mammals were named for Commerson, including Commerson's dolphin, from southern South America, with a subspecies known from the southern Indian Ocean near the Kerguelen Islands. No species resulting from the voyage were named for Baret. Commerson had planned to name a new genus for her, *Baretia* (now *Turraea*), but since he never formally described it in a publication, the name did not persist. It was not until 2012 that Baret's botanical contributions were recognized in *Solanum baretiae*, which was named for her by Eric Tepe, curator of the University of Cincinnati herbarium.

Baret and Commerson's systematic inventory and evaluation of the natural vegetation of Mauritius resulted in a report that recommended the preservation of natural flora. Even as early as the mid-1700s, tropical agricultural products (e.g., coffee, sugar, tea, spices) in high demand by those at home were replacing the island's forests. Poivre used this report to promote a law—issued in 1769!—requiring that steps be taken to inhibit or reverse the damages of past deforestation, including setting aside reserves in forests and wetlands, and replanting with native species. This law was one of the first governmental actions by a European country taken to conserve tropical ecosystems.

Beyond its scientific contributions, the Bougainville voyage contributed to the theory of utopian socialism, part of the philosophical backdrop to the French Revolution and later to manifestos written by Karl Marx and others. Bougainville gave effusive descriptions of Tahiti in his account of the expedition, describing it as an earthly paradise where men and women lived in blissful innocence, far from the corruption of civilization. While living in Mauritius, Commerson also waxed eloquent about life in Tahiti. In an article that appeared in *Le Mercure*, a French literary magazine, he wrote that the people of Tahiti were "born under the most beautiful of skies,

▲ ▲ *Bougainvillea spectabilis*, a common garden plant in warmer climates, was first collected by Baret and Commerson.

▲ Commerson's dolphin, *Cephalorhynchus commersonii*.

Type specimen of *Bougainvillea spectabilis*, collected by Baret and Commerson in Brazil.

TYPE   HERB. MUS. PARIS.

Herb. Muséum Paris

P00169376

Systematic Studies in Solanaceae

Solanum barctiae Tepe

Det. Sandra Knapp   BM   2016

287
8A

▲ *Te Arii Vahine* (King's Wife), Paul Gauguin, painted in Tahiti, 1896.

◄ A specimen of *Solanum baretiae*, a species in the same genus as tomatoes and potatoes.

fed on the fruits of a land that is fertile and requires no cultivations, ruled by the fathers of families rather than by kings; they know no other Gods than Love." These descriptions of Tahiti as a primitive utopia captured the imagination of Jean-Jacques Rousseau and other philosophers, who used them as the basis for a critique of the social hierarchy found in European society. Commerson later wrote a second letter that tempered his enthusiasm for Tahitian society a bit by pointing out some social problems on the island, but this more realistic account did little to quell the romantic concept of life there that had taken hold back home in France.

In response to Bougainville's expedition, England sent James Cook (1728–1779) on his first voyage around the world in 1768, just as the French expedition was making its way through the East Indies. Cook's voyage—with the botanical collecting team of Joseph Banks (1743–1820), a wealthy natural historian and patron of the

sciences, and Linnaeus student Daniel Solander—was considered a great success from the point of view of navigation and natural history specimen collection. Banks in particular was very fired up by the successful completion of the expedition and the accolades he received from it, which rather overshadowed the credit given to Cook's considerable navigational feats.

Soon after return, Banks began making plans for a second circumnavigation voyage to be led again by Cook, with an even larger scientific contingent. However, the final plan for the expedition approved by the Admiralty did not meet Banks' demands, and at the last minute he decided not to participate. The voyage went ahead anyway, leaving England in 1772. An odd incident early in the voyage was reminiscent of, or perhaps was inspired by, the Bougainville expedition. Cook's first stop was at Funchal, on Madeira, where a Mr. Burnett asked to join the expedition as a botanist to assist Banks in his explorations; upon hearing that Banks was not on board, however, Mr. Burnett hastily rescinded his request and left the ship. It was evident to Cook and others who met Burnett that he was in fact a woman in disguise. No further details of the story are known, but speculation was that she was one of Banks' lovers (he was said to have had quite a few) and that Banks had had a hand in organizing this subterfuge to take her along with him on the voyage:

HMS *Resolution* and
HMS *Discovery*, ships of
Captain Cook's second
circumnavigation, off Tahiti,
by William Hodges, 1776.

he had simply failed to contact her about his change of plans when he decided not
to go. Jeanne Baret, still in Mauritius at the time of this incident, surely never knew
of it, but it is interesting to contemplate what advice she would have given to Mr.
Burnett had she met her!

José Mutis and the Exploration of the New World

Although like other European powers Spain was very much engaged in the strategic colonization of faraway lands for military purposes and commerce, the country
followed a somewhat different model for natural history exploration. Spain had
secured a strong colonial dominance in the Americas early on, especially from Mexico southward, including permanent settlements and the establishment of European
institutions, and the Spanish government therefore invested in botanical exploration

not only for plants to bring home but also as a source of income to support colonial economies. The first major investment was the Botanical Expedition to the Viceroyalty of Peru, commissioned by Charles III and led by Hipólito Ruiz López, José Antonio Pavón, and French physician Joseph Dombey. The expedition began in 1777 and continued for about ten years, exploring what is present-day Peru and Chile; members collected plants for herbarium specimens and cultivation and made field illustrations as well. The expedition had many setbacks, including a fire in the building where specimens were stored in Macora, Peru, and shipwrecks, both during the course of the expedition and on the return trip. Still, 10,000 specimens representing more than 3,000 plant species made it back to Madrid's Royal Botanic Garden, where they are still housed. Ruiz and Pavón published the results of the expedition in *Florae Peruvianae et Chilensis Prodromus*, a richly illustrated ten-volume work.

José Celestino Mutis (1732–1808) was a Spanish botanical explorer who, unlike predecessors in this endeavor, pursued the work with a strong commitment to improving lives, not only in the country of his birth but in his adopted home, the colony of New Granada. New Granada was a vast territory including present-day Colombia, Ecuador, Panama, Venezuela, northern Brazil, and western Guyana.

Mutis was appointed physician to the viceroy of New Granada in 1760. A polymath with formal training in botany and medicine, Mutis was the most broadly trained of any explorer of South America of the period and immediately upon arrival began to develop a scientific infrastructure in the capital, Santa Fe de Bogotá. He was an early Spanish disciple of Linnaeus and, following Linnaean criteria, began classifying the local flora. He also sent specimens to Linnaeus and his son (also Carl Linnaeus), who, along with several other authors, described over 300 specimens based on his collections, including *Mutisia*, a genus of climbing daisies named for Mutis.

Mutis was very keen to develop plant study in New Granada, proposing the establishment of a botanic garden as well as expeditions to explore the entire

▲ José Celestino Mutis examining a red-flowered representative of *Mutisia*, a genus in the sunflower family, by R. Cristobal, 1930.

▶ Illustration of *Renealmia thyrsoidea* (as *Amomum thyrsoideum*) from *Florae Peruvianae et Chilensis Prodromus*, vol. 1, plate 2.

I. Galver del. et inc.

AMOMUM *thyrsoideum*.

territory. He spent almost 20 years trying to interest the Spanish government in funding such exploration and finally was successful. Given Mutis' reputation and the lack of knowledge about the plants of this vast, plant-rich area, the announcement of the Royal Botanical Expedition to New Granada caused great excitement in European naturalist circles. Mutis was appointed to the Royal Swedish Academy of Sciences and became a correspondent of the Royal Botanic Garden and a member of the Royal Academy of Medicine in Madrid; by including Mutis, these organizations could hope to receive periodic reports from him and learn of his most exciting finds in advance of published expedition results. The expedition launched in 1783 and continued for 34 years, the longest of any Spanish exploring expedition, covering more than 50,000 miles around the Magdalena River Valley in what is now central Colombia. The expedition was a collection of trips rather than a single prolonged one, and in addition to cataloging the flora, it was also charged with developing a mining industry and developing new agricultural crops for export.

The effort was based first in the town of Mariquita, near the Magdalena, where Mutis founded a botanic garden; in 1791, the headquarters moved to Bogotá, where the Casa de Botánica functioned like a scientific institute, employing some 35 local people, including artists and plant collectors. The artists Mutis trained were mostly mestizos, and they developed a distinctive illustration technique: they used colors derived from native vegetation that were far more vibrant, especially in reds and greens, than those used in European plant illustrations of the time. Plates composed by Salvador Rizo, Francisco Javier Matis, and other masters of the style included a central drawing of the plant showing its leaves and flowers, surrounded by smaller drawings, sometimes not colored, of the key features that distinguish the species.

Mutis' botanical expedition took place during a time of political turmoil in New Granada. Inspired by the French Revolution, citizens of New Granada wanted autonomy from Spain, resulting in Colombia's declaration of independence in 1810. The botanical expedition team included several men who would become important figures in the Colombian revolution, namely Francisco José de Caldas, Antonio Nariño, and Francisco Antonio Zea, some of whom, such as Caldas, would later be executed for their role in the uprising. Mutis' nephew, Sinforoso Mutis, was also a revolutionary; he continued the botanical exploration for a time after the death of his uncle in 1808. It isn't clear that Mutis intentionally created an atmosphere that was conducive to insurgents, or whether these ideas were naturally prevalent among the well-educated young men who would have been drawn to such work.

Pablo Morillo, leader of the Spanish troops sent to quell the uprisings in New Granada, demanded the return of the expedition specimens and drawings to Spain. Mutis had intended that these resources remain in Bogotá as part of the nascent

L. 21.

Mutisia clematis.

Mutisia clematis
by Salvador Rizo.

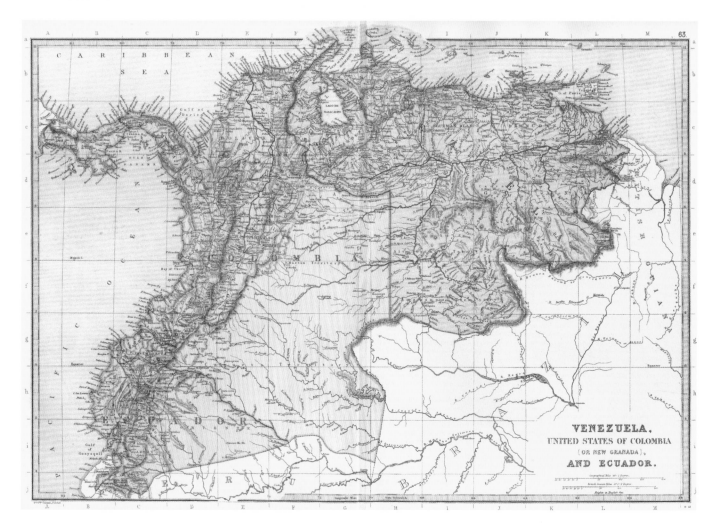

scientific infrastructure there, but the Spanish government insisted that the expedition materials were critical to studies by botanists at Madrid's Royal Botanic Garden and were unsafe in Bogotá due to the political unrest. Sinforoso Mutis was arrested in 1814, but while in prison, he was taken daily to the Casa de Botánica to prepare the approximately 24,000 expedition specimens and notes for return to Spain. Once in Madrid, the specimens were packed away and forgotten, suggesting that the motivation for their return had less to do with protecting them than with hindering the realization of Mutis' vision for his adopted home.

For generations, botanists believed that the Mutis specimens had been destroyed. However, Smithsonian botanist and South American explorer Ellsworth Killip located them at Madrid's Royal Botanic Garden in 1929. The specimens, untouched for more than a century, were mostly well preserved. Killip organized the specimens, and since many included duplicates, he was able to keep a set for his own institution, the U.S. National Herbarium at the Smithsonian. Duplicates were made available

Map of the Viceroyalty of New Granada published in the *Descriptive Atlas of the World and General Geography* by Blackie & Son, London, 1893.

In the image, the following labels appear:

Herbario de Mutis
MA 662097

Real Jardín Botánico. C. S. I. C.
HERBARIO NUMERADO EN 2008

HERB.HORT.REG.MATRIT

662097

MA

HERBARIUM HORTI BOTANICI MATRITENSIS

Plantae Expeditionis Botanicae Mutisii
Vice-Regni Novae-Granatae.
(1783-1808)

Mutisia clematis L.f.

det. S. Díaz P. 1985

NÚM. 5911

Type specimen of *Mutisia
clematis*, from Mutis'
herbarium, Royal Botanic
Garden, Madrid.

to many other herbaria as well. The original set of specimens, still held at the herbarium in Madrid, includes 6,717 drawings and 20,000 specimens representing 6,000 species. It was not until the mid-20th century that Spain and Colombia joined forces to publish Mutis' illustrations and scientific descriptions from the Royal Botanical Expedition; in total, 23 volumes of *Flora de la Real Expedición Botánico del Nuevo Reino de Granada* have been published so far, the first in 1954.

Mutis was the first to discover quinine (aka *quina*) trees in Colombia. Used to treat malaria, the plant source for this medicine is the bark of several species in *Cinchona*, a genus in the coffee family. It was one of the first important medicinal plants brought from South America back to Europe in the 17th century. Mutis carried out a series of scientific studies on *Cinchona*, which were published as newspaper articles; his "El Arcano de la Quina" (The Secrets of Quina) appeared in the *Papel Periódico de Santafé de Bogotá*, a small-format weekly newspaper, in 1793 and 1794. Here he recorded everything he knew about the genus and illustrated it as well. The newspaper was Mutis' preferred publication outlet; he described several plants there that his nephew later published in a more conventional scientific outlet.

In the final years of his life, Mutis pursued a vigorous program of civic improvement and institution building in Bogotá. These activities included running a vaccination campaign for citizens and overseeing the construction of the first observatory and one of the largest scientific libraries in the New World. His observatory still exists, a national landmark valued for its architecture and historical significance: Colombia's independence movement leaders met there to plan their revolt against Spanish rule. Mutis developed plans for a botanic garden in Bogotá that was never built; however,

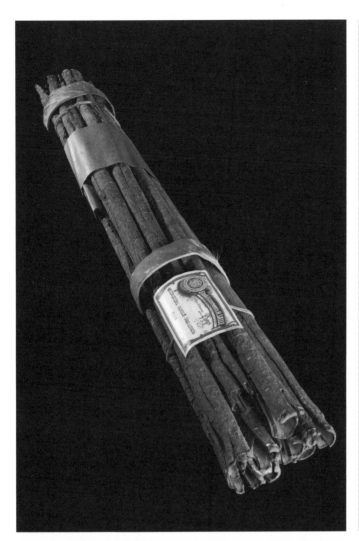

▲ Specimen of *Cinchona officinalis* (Peruvian bark).

▼ Cinchona bark packaged for sale by F. Laurent, a pharmacy in Paris, France, in about 1930.

◄ Mutis is featured on Colombia's 200 pesos note (issued in 1991, now discontinued), one of the few scientists to be so honored. The observatory is visible in the background.

Astronomical observatory in Bogotá, the construction of which was overseen by Mutis.

in 1955, the city finally created the José Celestino Mutis Botanical Garden, which contains an herbarium. Unlike most European explorers of his time, Mutis' most immediate contribution was to his adopted homeland, where he built a foundation for scientific education and research, and in the process either actively or passively promoted national independence. Though well known and respected for his botanical work during his lifetime, it would be many years until the full impact of his botanical work would be recognized in Europe. The rediscovery of his herbarium helped to set in motion a renewed effort to document the flora of one of the most mega-diverse countries in the world, work that continues today.

Richard Spruce, Explorer and Scholar

Scientific discovery as the main objective of an exploring expedition became commonplace in the 19th century, leading to more far-reaching and well-documented results. Alexander von Humboldt was a German aristocrat and naturalist who traveled to South America with French botanist Aimé Bonpland from 1799 to 1804, where they explored more than 9,000 miles in New Granada and other Spanish territories in what are now Cuba, Mexico, and Peru. They stayed several months with Mutis in Bogotá. Humboldt and Bonpland made about 5,000 botanical collections, the largest set of which is held at the National Herbarium in Paris but with others widely dispersed among European herbaria and represented in several North American herbaria as well. A keen observer and disciplined journal-keeper, Humboldt's publications were as important as his specimens. In his *Essay on the Geography of Plants* (1807), he related the plant communities he had encountered to local temperature and other environmental attributes, representing early contributions to the disciplines of ecology and biogeography. His multi-volume *Cosmos* (1845–58, the last two volumes published posthumously) represents an attempt to unify biology and geography into a synthesis of the world's environment. These

works were widely read and inspired the generation of explorers who followed, including Charles Darwin.

    A young British aristocrat who had just completed a degree at Cambridge in theology, Darwin was offered the opportunity to sail on a scientific expedition led by Robert FitzRoy as the captain's traveling companion. The expedition, aboard HMS *Beagle*, left England in 1831, traveled around South America (stopping famously for five weeks at the Galapagos Islands, visited a century earlier by William Dampier), crossed the Pacific to New Zealand and Australia, and then returned home around the Cape of Good Hope, completing the circumnavigation in 1836. Because Darwin was not primarily responsible for the scientific results of the expedition, he was freed from the day-to-day grind of collecting data and managing collections, allowing him more time to observe and reflect upon the biodiversity and landforms he encountered. The book that resulted from this trip, *On the Origin of Species* (1859), led to the theory of natural selection that provided a testable hypothesis for the multiplicity

Depiction of Alexander von Humboldt and Aimé Bonpland at the foot of Chimborazo, a volcano in Ecuador, by Friedrich Weitsch, 1806.

Specimen of the liverwort *Lepidozia chordulifera*, collected by Charles Darwin on Chile's Chonos Archipelago. The specimen, which came from William Hooker to William Mitten, was taken out of the small blue packet and placed inside the circular container for this photo. The illustration is from the original publication by Thomas Taylor in the *London Journal of Botany* in 1846.

of lifeforms that he and all explorers before him had encountered. Although not a major botanical collector, Darwin did make about 1,100 herbarium specimens, which he gave to his mentor, John Henslow of Cambridge University, where they are still maintained. Henslow and others described about 115 new species of plants and fungi based on Darwin's collections.

Following Humboldt and Darwin was Richard Spruce, who is far less famous but who matched his predecessors step for step in tenacity in the field and scholarship at home upon return. Spruce was born in 1817 in Ganthorpe, near Castle Howard, in Yorkshire. Son of a local schoolmaster, he began his career as a teacher as well. Spruce had a chronic bronchial condition since childhood, the prescribed treatment for which was spending as much time as possible in the fresh air, which may have led to his interest in botany and his passion for collecting and documenting plants. He was especially interested in bryophytes, particularly hepatics, or liverworts. He became known to the British botanical community through the observations he published on the mosses and liverworts of Yorkshire in *The Phytologist*, a popular botanical journal.

When Spruce was advised to visit a warmer climate for his health, William J. Hooker, director of the Royal Botanic Gardens at Kew, suggested a collecting trip to the Pyrenees. Spruce spent a year there (1845–46), financing the trip through the sale of his bryophyte specimens. By the mid-19th century it had become common for collectors to finance their trips by selling specimens to subscribers, who included amateur botanists as well as directors of institutional herbaria. The practice was far more common for cryptogams, which required specialized knowledge and sometimes specialized collecting techniques. The collector would make sets of duplicate specimens, or exsiccatae (sing., exsiccata; from the Latin *exsiccare*, "to dry"). Although sometimes used as general terms for any dried plant specimen, these words have a special meaning in botany, referring to one or more sets of specimens distributed to represent a particular type of organism or geographic area. Exsiccatae were never standardized through formalized rules; however, by a more or less general understanding, specimens were numbered sequentially, with each number representing a single gathering from which duplicate specimens were prepared, so that each set would contain specimens from the same gatherings. Some exsiccatae were issued as a single set; others were issued in multiple sets over a period of time.

Immediately upon his return, Spruce produced a major publication on the mosses and hepatics of the Pyrenees, earning him a reputation as a keen observer and a first-rate collector of all types of plants, and he was eager for more fieldwork. In 1849 he left for South America, again encouraged by Hooker and with Hooker's colleague at Kew, botanist George Bentham, acting as his broker. In this role, Bentham received specimens from Spruce, identified them or distributed them to specialists for identification, and then offered them for sale on Spruce's behalf. Spruce had 20 subscribers for his South American specimens at first, but as the high quality of the specimens became known, the total number of subscribers nearly doubled. We do not know how much Spruce earned from the sale of specimens.

Spruce started his journey from Belém, at the mouth of the Amazon. From there he traveled to Santarém, at the mouth of the Tapajós River, where he met up and traveled for a time with Henry Bates and Alfred Russel Wallace, fellow countrymen who were engaged mostly in zoological collecting. Bates would spend a total of 11 years in South America, primarily collecting insects; Wallace stayed about four years in South America and then later traveled to the Asian tropics, where after eight years of study, he came to conclusions concerning the origin of species that were very similar to those of Charles Darwin. His observations about the discontinuity in species of plants and animals between the islands of tropical Asia and those of Australasia, a phenomenon that has come to be known as Wallace's Line, contributed to the understanding of the movement of continental plates in the geological history of the region.

After his time with Wallace and Bates, Spruce continued on alone, exploring along the Amazon and its tributaries for the next several years. He then traveled up the Amazon to Peru, ending in Tarapoto on the eastern slope of the Andes; he stayed there two years before moving on to Ecuador, where he collected in the vicinity of Baños, at the foot of Tungurahua, and after that near Ambato. While in Ecuador he

▲ Richard Spruce sat for this photograph shortly after his return from his expedition to South America.

▶ Diversity of hepatics illustrated by Ernst Haeckel (1834–1919), vol. 1, plate 82, from his *Kunstformen der Natur* (Art Forms in Nature), published in 1904

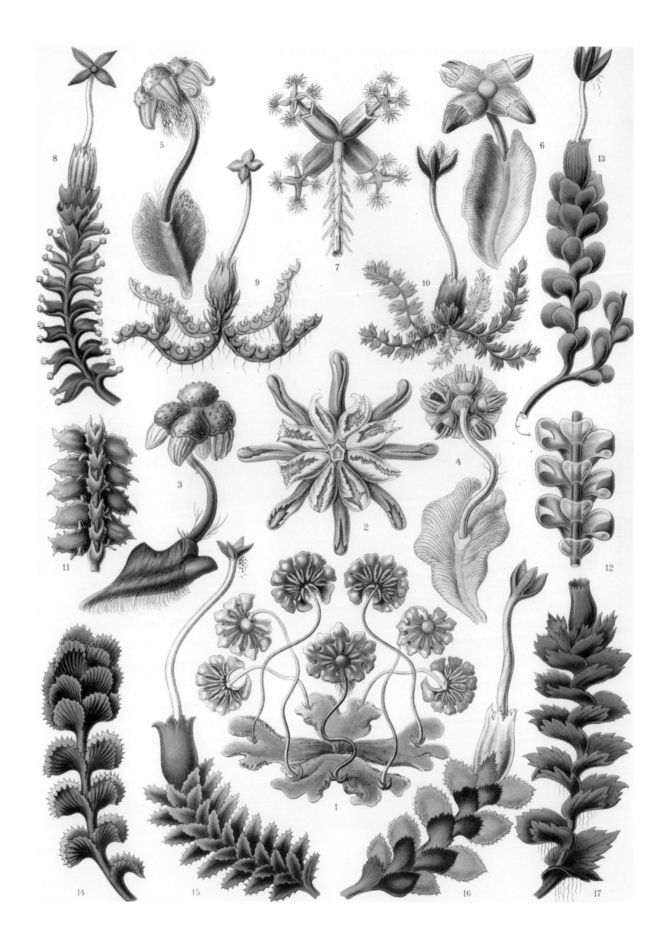

was hired to collect specimens of *Cinchona* species to be sent to India for planting there. Spruce collected (or supervised the collecting of) 100,000 seeds and 600 cinchona plants on the western slopes of Chimborazo. He brought the samples to Guyaquil, where he handed them off for shipment to India and received his payment. Although clearly an act of biopiracy, this was not considered a crime at the time, and since Spruce's plants became the foundation of plantations in India and Ceylon, his actions brought relief to thousands of malaria sufferers in that region. He published a report on this undertaking in 1862.

Spruce invested all his savings and earnings from the sale of his cinchona plants in a mercantile house in Guayaquil, but it soon went bankrupt. And his health, never robust, worsened during his years in South America. He suffered through bouts of malaria and other maladies, and while in the Andes he developed a condition that worsened to a point where later in his life he could hardly bear the pain of sitting upright. He had chronic diarrhea, severe headaches, and frequent violent coughing fits. Sick and penniless, Spruce returned home to England in 1864.

Spruce's 15 years of plant collecting resulted in more than 30,000 specimens. In contrast to earlier explorers, Spruce's collection notes and specimen labels are considered nearly flawless examples of botanical record keeping. His handwriting was consistently legible, and his notes included precise locations and dates. Bentham stewarded his collections well, studying many of his specimens of flowering plants himself and distributing the ferns, lichens, and fungi to other British specialists. When he first returned to England, Spruce lived near London, in Hurstpierpont, Sussex, to be close to his colleague William Mitten, who had agreed to write up Spruce's mosses; the resulting book, *Musci Austro-Americani*, was published in 1869. Mitten, who earned his living as an apothecary, also provided Spruce with medicines,

▲ Plaque from Spruce's cottage in Coneysthorpe, where he lived after his return to Yorkshire.

▼ Richard Spruce's cinchona plants, upon their arrival in India.

Specimen of the palm *Lepidocaryum tenue* var. *casiquiarense*, collected (as *L. casiquiarense*) by Richard Spruce along the Casiquiare River, which branches off the Orinoco River in Venezuela and empties into the Rio Negro, a major tributary of the Amazon.

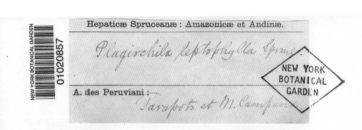

Hepaticæ Spruceanæ : Amazonicæ et Andinæ.

Plagiochila leptophylla Spruce

NEW YORK
BOTANICAL
GARDEN

A. des Peruviani :—

Tarapoto et M. Campana

Plagiochila leptophylla Spruce TYPE
Trans. & Proc. Bot. Soc. Edinburgh
15:475, 1885

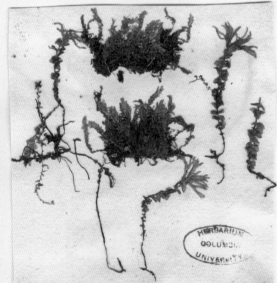

P. leptophylla S.

most of which were ineffective at giving more than temporary relief from his maladies. A few years later, Spruce moved back to Yorkshire, settling in Coneysthorpe, not far from his birthplace, where he remained for the rest of his life.

Spruce saved the palms and the hepatics he collected for his own study. For the palms he reviewed not only his own collections but also other Amazonian collections stored at Kew that Hooker sent to him for study. The resulting work, published in 1871, included detailed descriptions of the key distinguishing features of each species supplemented with his field observations. Spruce's magnum opus was *The Hepaticae of the Amazon and the Andes of Peru and Ecuador*. Now over 130 years old, it is still an important reference on tropical South American hepatics. The work, which contained almost 400 new species and several new genera, provided a new classification for the group influenced by Darwin's theory of natural selection. The study of hepatics is painstaking because the plants are usually very small, requiring the use of a microscope to describe key features. In continuing poor health, Spruce had to work mostly while reclining on a bed or couch and could often spend no more than a few minutes at a time upright at the microscope.

Additionally, Spruce recorded uses of plants by indigenous people during his travels, trying to determine the botanical sources of the various gums, fibers, resins, oil, dyes, drugs, narcotics, and foods he encountered. He provided science with some of the first extensive botanical knowledge of *Hevea*, the source of rubber. Furthermore, he recorded the customs and traditions of the various communities he met, making him one of the earliest ethnographers in the Amazon and Andes. He created vocabularies for 21 Amazonian languages and made maps of three previously unexplored rivers. His specimens are fairly widely distributed, with major sets of flowering plants and ethnobotanical samples in the herbaria at the Royal Botanic Gardens, Kew, the Royal Botanic Garden Edinburgh, the Natural History Museum in London, the Manchester Museum, and Trinity College in Dublin. The most complete set of Spruce's mosses are deposited in William Mitten's herbarium at the New York Botanical Garden. Spruce is commemorated in the names of many plants, including the moss genus *Sprucella*, the hepatic genus *Sprucea*, and several orchid species.

After his death, Spruce's travel notebooks were edited by Wallace and published as *Notes of a Botanist on the Amazon and Andes* (1908), a work filled with fascinating insights into biology and human culture. Unlike Humboldt, Darwin, and Wallace, Spruce focused on documenting the subtle details of nature rather than the broad patterns of diversity, and although this approach may have kept him just outside the inner circle of famous explorers, it clearly had its own rewards. Spruce was steadfastly unapologetic about the small plants to which he was devoted, writing:

◄ Type specimen of the hepatic *Plagiochila macrostachya*, collected (as *P. leptophylla*) by Richard Spruce, from the herbarium of William Mitten, on deposit at the New York Botanical Garden.

I like to look on plants as sentient beings . . . which live and enjoy their lives—which beautify the earth during life, and after death may adorn my herbarium . . . . It is true that the Hepaticae have hardly as yet yielded any substance to man capable of stupefying him, or of forcing his stomach to empty its contents, nor are they good for food; but if man cannot torture them to his uses or abuse, they are infinitely useful where God has placed them, as I hope to live to show; and they are, at the least, useful to, and beautiful in, themselves—surely the primary motive for every individual existence.

Later Development of European Herbaria

In 1700, Europe boasted only a small number of institutional herbaria, the majority held in bound volumes; and collectively, they contained perhaps several hundred thousand specimens, the majority collected in Europe. By the end of the 19th century, however, botanists had enough specimens at their disposal to summarize the major plant groups of the world. These specimens were key to the monumental publication *Die Natürlichen Pflanzenfamilien* by German botanists Adolf Engler and Karl Anton Prantl, published in 23 volumes between 1887 and 1915. This work characterized all known plant genera grouped by family, following a novel classification system. Though now superseded, the Englerian system was used by botanists for many years, and the arrangement of some herbaria still follows it, because of its comprehensiveness.

By 1900, there were about 250 herbaria in Europe, held in 18 countries. Another 368 herbaria would be formed in the 20th and early years of the 21st centuries, but of the nearly 1,100 herbaria once listed in Europe, about a third no longer exist. Many smaller herbaria were combined with larger ones, but the status of some is unknown.

One institution notably missing from a list of the largest herbaria in present-day Europe is the herbarium of the Berlin Botanical Garden. At the beginning of the 1940s, it contained about 4 million specimens, built over 175 years, making it one of the largest in the world. After World War II began, some precautionary measures were taken to safeguard the collection in case of a direct hit from Allied bombing. Specimens of algae, fungi, and bryophytes that had been stored in the attic were moved to a safer location in the basement, and the herbarium of Carl Ludwig Willdenow, botanical mentor of Alexander von Humboldt, with more than 20,000 type specimens, was evacuated to a mine shaft near Bleicherode in central Germany. The herbarium staff was in the process of preparing more collections for storage offsite when an Allied bomb hit the eastern part of the building, which housed the herbarium and library, on the night of 1 March 1943; most of the herbarium's collections

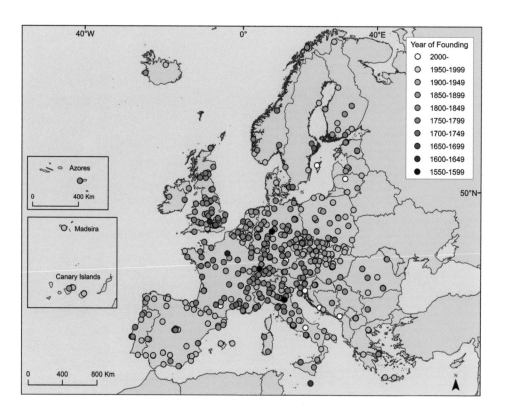

Map showing location of European herbaria, points color-coded by year of founding.

were lost to fire or water damage, and about 30,000 specimens that were on loan from other herbaria were destroyed as well. As Elmer Merrill wrote in the December 1943 issue of *Science* magazine, "The loss of the Berlin herbarium is a catastrophe of major proportions to world botany." Only about half a million of the 4 million total specimens remained after the war. The restoration of the building was completed in the mid-1950s, and the rebuilding of the herbarium began, thanks to donated specimens from herbaria throughout Europe and North America. In 1979, the herbarium moved to a new wing, where collection storage is underground. Berlin's herbarium now houses about 3.8 million specimens, nearly back to its pre-destruction accession total. Furthermore, in 1929, James F. Macbride of Chicago's Field Museum had obtained funding from the Rockefeller Foundation to photograph European type specimens as a reference for American botanists who did not have the means to travel to Europe. Approximately 40,000 type photographs were captured through this project, including many of those lost in the bombing of Berlin, so there is a photographic record of at least some of the lost Berlin specimens.

The approximately 800 herbaria now active in Europe hold approximately 175 million specimens, about 46% of all known herbarium specimens. About 45% of European herbaria are associated with universities, 38% with botanic gardens and museums, and 17% with large federally funded museums or research institutes.

Washington Co., Utah, USA

Adiantum pedatum L.

Oak-pinyon-juniper comm., on basa
at ca 5150 ft elev.

Development of Herbaria in the United States

Interest in the plants of North America began for Europeans with the plants introduced by Columbus. Exploration of this new land for sources of food, medicine, and timber began shortly thereafter, along with the collection of plants for ornamental uses. Scientific interest in the plants followed directly, but in the early years, collections and observations of New World plants and fungi were made solely to benefit the scientific communities of the colonial powers.

The first herbarium specimens collected in North America are probably those of Michel Sarrazin, a physician living in Quebec. Beginning in 1697 he sent specimens annually to French botanist Joseph Pitton de Tournefort, including the first known specimens of *Aralia nudicaulis* (ginseng) and *Sarracenia purpurea* (common pitcherplant). These specimens are maintained in the National Herbarium in Paris.

◀ An herbarium specimen of *Adiantum pedatum* (northern maidenhair fern).

◄◄ *Sarracenia purpurea*, one of the first plants collected in North America, by Michel Sarrazin (1659–1734), king's physician for New France (Quebec).

◄ *Sarracenia*, a genus of insectivorous plants native to wet habitats in North America, is named for Sarrazin. Species attract insect prey with secretions near the lip of their "pitcher" leaf; insects slip off the rim into the pitcher, where they are digested by enzymes in the liquid at the pitcher's base.

► Georg Dionysius Ehret's illustration of *Magnolia grandiflora* (southern magnolia) from Catesby's *Natural History of Carolina, Florida and the Bahama Islands*. The work contains 220 colored plates.

►► Specimen from a Sloane Herbarium volume, probably collected by John Banister, the first person known to have made herbarium specimens in the United States.

In what is now the United States, the earliest botanical activity focused on the southeastern portion of the country. The first specimens were probably those collected by John Banister (c.1649–1692), an Oxford-educated chaplain who settled in 1679 near the mouth of Virginia's Appomattox River. He established a parish there and took an interest in documenting the plants of the region, sending specimens and seeds back to England. John Ray included a list of Banister's plants in his *Historia Plantarum*, the same publication that included the species collected by William Dampier. Banister's botanical career was short—he was accidentally shot and killed by a field companion while collecting. Most of his specimens are in the Sloane Herbarium of the Natural History Museum in London, though most are not explicitly labeled as having been collected by Banister.

English naturalist Mark Catesby (1683–1749) visited Virginia, North and South Carolina, and the Caribbean on several trips between 1712 and 1726, making collections and illustrations that formed the basis of his *Natural History of Carolina, Florida and the Bahama Islands* (1730–47). Catesby's plant specimens, now preserved at both the Sloane Herbarium and Oxford University, have been imaged and are available online.

During his travels in America, Catesby met John Clayton (c.1694–1773) and taught him how to make botanical specimens. Clayton was an English immigrant

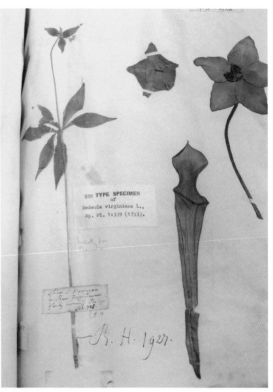

who served as county clerk of Glouces-
ter Country, Virginia, for many years
and explored rather widely in his free
time. His specimens found their way to
Dutch botanist Jan Gronovius, whose
Flora Virginica, published between
1739 and 1743, was based on collections
made by Clayton. Gronovius enlisted
help from Linnaeus in studying Clayton's specimens, and Linnaeus named
the lovely wildflower *Claytonia virginica* (spring beauty) for him. After
Clayton's death, his herbarium of some 700 specimens was purchased by Joseph
Banks; it is maintained as a separate collection at the Sloane Herbarium, with some
specimens also in the Linnean Herbarium.

Linnaeus convinced the Swedish government to fund a collecting trip to North
America by Pehr Kalm, with the primary purpose of obtaining living specimens of
mulberries adapted to the north temperate zone. In warmer areas, species of mulberry
(*Morus*) support silk-producing caterpillars, and the Swedish government hoped

◄ Clayton specimen of *Parthenium integrifolium* (wild quinine) from Gronovius' herbarium. Note the embellishment of the herbarium sheet with an urn for the specimen, similar to those used by Clifford.

► *Claytonia virginica* (spring beauty). Linnaeus named the genus for John Clayton.

▲ Specimen of *Quercus palustris* from the Michaux collection at the National Herbarium in Paris.

to adapt this industry to a colder climate. A Swedish silk industry did not develop from this effort, but Linnaeus based 60 new species on the plants Kalm collected in what is now the northeastern United States and southern Canada between 1748 and 1751. Kalm's specimens are part of the Linnean Herbarium.

In 1785, on the orders of Louis XIV, French botanist and explorer André Michaux (1746–1802) came to North America with his son François (1770–1855) to search for plants that could be useful for France agriculturally and as building materials and medicines. They established botanic gardens first near Hackensack, New Jersey, and later near Charleston, South Carolina. The Michauxs lost their government support during the French Revolution, but they continued to collect, exploring from Philadelphia across the Alleghenies as far as the Mississippi River. André was Thomas Jefferson's first choice as botanist on what would become the Lewis and Clark Expedition, but Jefferson came to suspect that Michaux was a spy for the French government and withdrew the offer. Between 1785 and 1793 André shipped 90 cases of specimens, living plants, and seeds back to France, where they are maintained as a separate historic collection (2,192 specimens) at the National Herbarium in Paris. He described many species in two works, *Histoire des Chênes de l'Amérique* (History of American Oaks) published in 1801, and *Flora Boreali Americana* (Flora of North America), published posthumously in 1803, both sumptuously illustrated by renowned Flemish botanical artist Pierre-Joseph Redouté.

François maintained the garden at Charleston, South Carolina, for a time but returned to France in 1803, where he wrote his *North American Sylva* (1817–19),

Pl. 27.

P.J.Redouté del.

Gabriel Sc.

Pin Oak.

Quercus palustris.

◄ *Quercus palustris*
(pin oak) plate by Redouté
from *Histoire des Chênes
de l'Amérique* (1801).

► John Bartram,
"The Botanist," by noted
American illustrator
Howard Pyle, 1879.

also illustrated by Redouté. Even after his return to France, François continued to support the advancement of botanical knowledge in Charleston, providing relevant literature to botanist Stephen Elliott (1771–1830), a plantation owner and legislator. Elliott's *A Sketch of the Botany of South Carolina and Georgia* (1816–24) was the most thorough scientific work focused on southern plants at the time. Elliott's herbarium, one of the earliest in the United States, is preserved in the Charleston Museum's herbarium, with some duplicates in the Torrey Herbarium at the New York Botanical Garden.

The Botanist.

John Bartram: A Foot in Both Worlds

Capital of the United States until 1800, Philadelphia was also the young country's academic center. The city was home to the American Philosophical Society, the nation's first scholarly organization, founded in 1743. The Academy of Natural Sciences, the first organization dedicated to the study of science in the United States, was founded in 1812. Philadelphia and the surrounding countryside was also an early center for the study of American plants and for building resources for that work, such as herbaria.

John Bartram (1699–1777) was one of the founders of the American Philosophical Society. Born into the Quaker community in Philadelphia, he was a farmer and nurseryman who received funding from England for extensive plant exploration, making him a transitional figure in the establishment of an American botanical tradition. With his son William he collected plants from New England west to Lake Ontario and south as far as Florida, traveling on horseback. In southern Georgia they made the first collection of *Franklinia alatamaha*, a beautiful flowering tree that he named for his friend Benjamin Franklin in *Arbustrum Americanum* (1785), authored by Bartram's cousin Humphry Marshall. This plant was collected only a few times after the Bartrams discovered it, and never again after 1830.

Bartram sent all his specimens and seeds back to England, where today they are held either at the Sloane Herbarium or the Linnean Herbarium. The only Bartram collections in the

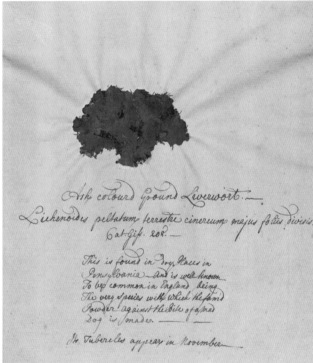

United States are maintained at the Sutro Library on the campus of San Francisco State University. Adolph Sutro was a German immigrant who made his fortune in silver from the Comstock Lode in Nevada and later became mayor of San Francisco. He used some of his wealth to establish an extensive library, including the herbarium of Robert James, 8th Baron Petre of Thorndon Hall, Essex, England, an avid horticulturalist who helped support the Bartrams' collecting trips. Their specimens are held in two of the 17 volumes that make up Lord Petre's herbarium.

▲ One of John Bartram's specimens from Lord Petre's herbarium. This "Ash colourd Ground Liverwort" is actually a lichen, *Peltigera canina*, common in Europe and North America and "the very species with which the fam'd Powder against the Bite of a mad Dog is made" (hence the specific epithet).

▼ Though apparently extinct in the wild, *Franklinia alatamaha* is grown extensively in cultivation.

Muhlenberg and Schweinitz: Pennsylvania Pastor-Botanists

Gotthilf Henry Muhlenberg (1753–1815) and Lewis David de Schweinitz (1780–1834) were both clergymen born in America but educated in Germany. In 1813, Muhlenberg made his first solo publication, an account of all the known plant species growing near his home in Lancaster, Pennsylvania. His most ambitious work, published posthumously, was his treatment of grasses of North America, which included not just lists of plant names but characterizations of individual species based on his own examinations of them.

Lewis David de Schweinitz was a member of a prominent family in the Moravian church, an offshoot of Lutheranism, and the first to document the fungi of North

▲ Bartram's Garden in Philadelphia is now a National Historic Landmark.

◀ Gotthilf Henry Muhlenberg (1753–1815).

▶ *Muhlenbergia rigens* (deergrass), one of many muhlys in this popular genus of ornamental grasses.

America in a scientific manner. He was also a child prodigy, writing his first botanical work, on the local flora near his home in Bethlehem, Pennsylvania, at age ten. He was educated at the Moravian Theological Institute in Silesia (now far-eastern Germany), where he studied with theologian and mycologist Johannes Baptista von Albertini. Upon his return to America, he began his mycological and theological careers in Salem, North Carolina. His first scientific work, *Synopsis Fungorum Carolinae Superioris*, was published in Europe with the assistance of German mycologist Christian Schwaegrichen in 1822. He later returned to Pennsylvania, where he wrote two additional compendia on North American fungi that included his own skillfully executed paintings. In total, his three works on fungi contain about 4,500 species, 1,500 of which were new. Many species were named for him, including *Phaeolus schweinitzii* (dyer's polypore). Schweinitz's home in Bethlehem is maintained as a National Historic Landmark.

The published literature available for identifying American plants and fungi being very limited at the time, Muhlenberg and Schweinitz each built an herbarium to help them identify the plants and fungi they encountered. Their herbaria grew

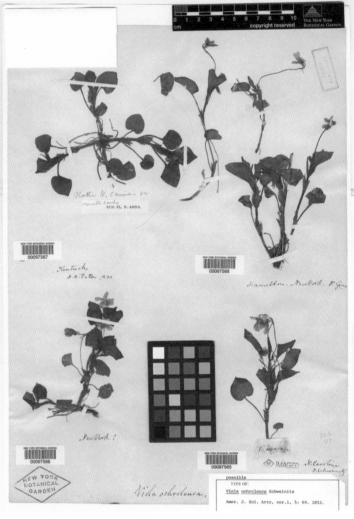

▲ Type specimen of *Viola ochroleuca* collected by Schweinitz.

▼ Lewis David de Schweinitz (1780–1834).

not only through their own collecting but also through the exchange of specimens with a large network of correspondents that included European botanists as well as other American naturalists. Muhlenberg's herbarium at the time of his death consisted of approximately 17,000 plant specimens, including 1,500 packets of algae, bryophytes, and lichens. The Schweinitz herbarium, the largest private herbarium in the New World at the time, consisted of 23,000 specimens of fungi, bryophytes, and flowering plants. These herbaria, both maintained at the Academy of Natural Sciences in Philadelphia, preserve a record of how floristic knowledge grew from the ground up in this new nation.

Muhlenberg and Schweinitz, and their frequent correspondent Stephen Elliott, approached the study of fungi and plants primarily on a local level, focusing on the organisms with which they had personal experience. They reasoned that regional knowledge such as theirs and others like them would eventually be knit together to form a flora of the entire country. They shared their experiences and specimens (usually small bits of them), through correspondence. Their letters to one another are sometimes attached to specimen sheets (often in an envelope or packet) in their herbaria or are preserved in library archives of the American Philosophical Society.

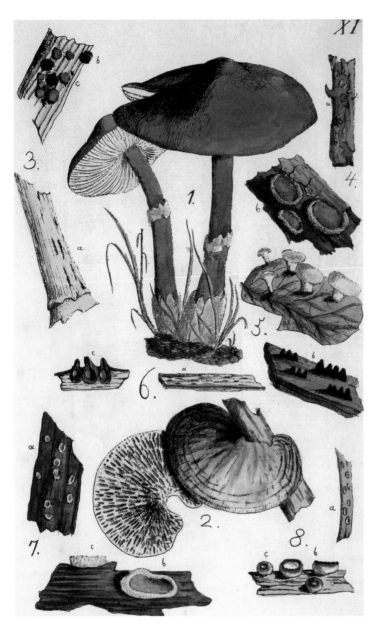

XI

▲ Plate 11, from Albertini and Schweinitz's *Conspectus fungorum in Lusatiae superioris agro Niskiensi crescentium* (1805).

The Twisted Tale of Lewis and Clark's Botanical Specimens

Benjamin Smith Barton (1766–1815), a Philadelphia physician, academician, and botanist, envisioned a different approach to documenting the American flora. In an address to the Philadelphia Linnean Society in 1807 he presented a sweeping vision for future plant research that would establish the continent-wide distributions of key American plants, especially those used for timber and other products, and would track changes in the flowering and fruiting times of plants over the years as a way to understand climatic change, an idea that remains under active investigation. As resources for this work, he called for a comprehensive herbarium of American plants and a national flora that would surpass the contributions of the Michauxs. Barton made little actual progress toward realizing his vision, but his grand plan attracted the attention of Thomas Jefferson, who as president enlisted Barton's help in establishing the methods and goals for botanical exploration on the Lewis and Clark expedition, the planning for which began when Philadelphia was still the nation's capital.

Led by Meriwether Lewis and William Clark, the expedition team departed in 1804. Over the course of the next two years they explored the region between St. Louis and the Pacific Coast, which they reached at Fort Clatsop near present-day Astoria, Oregon. The scope and competent execution of the expedition, which took place during a relatively peaceful interlude between the Revolutionary War and the War of 1812, captured the American imagination. Americans read the detailed accounts from the expedition of unknown lands newly acquired by the United States through the Louisiana Purchase with great interest.

Thanks to very specific instructions from Jefferson, himself a noted data-recorder, the journey was admirably documented with journal entries, sketches, and both biological and ethnographic collections. Barton trained Lewis in botanical collecting techniques and provided him with reference works, including Linnaeus' *Systema Naturae*, to take along on the trip. Though apparently neither Lewis nor Clark had prior experience with botanical specimen preparation, they carried out this work diligently throughout the expedition.

Lewis and Clark's care in documenting their specimens stands in contrast to the way the collections were handled after the expedition. The story of Lewis and Clark's botanical specimens is complicated and controversial—and contains some mysteries that still have not been solved. The first set of their plants reached Jefferson in 1805. These had been collected on the first leg of the journey between St. Louis and Fort Mandan, their overwintering camp. Jefferson promptly sent the specimens to the American Philosophical Society, who referred them to Barton for identification and description.

Barton did nothing with the specimens. Despite his grand botanical ambitions for the nation, Barton had many other commitments as professor and physician and probably not much actual experience identifying and describing species, especially those from a previously unexplored area. A second set of Lewis and Clark plant collections, part of a cache stored near Great Falls, Montana, before the ascent of the Continental Divide, was lost due to flooding during the winter of 1805.

Upon the expedition's return, Lewis brought the remaining botanical specimens to Philadelphia. On Barton's recommendation, Lewis hired Frederick Pursh (1774–1820) to identify, describe, and illustrate them. Pursh was a German botanist and nurseryman who came to the United States in 1799. He became part of the Philadelphia botanical community, trained and supported by Barton and others.

A Lewis and Clark specimen, collected in 1805. Although the handwritten labels at the top of the sheet refer to *"Lewisia nervosa,"* the plant was eventually described as *Berberis nervosa* (dwarf Oregon grape), a member of the barberry family.

Pursh studied the specimens during the winter of 1807–08, producing a set of draft descriptions and some illustrations. He tried repeatedly to contact Lewis for further direction on the work but heard nothing. By this time Lewis, who had been appointed governor of the upper Louisiana territory, was heavily preoccupied with management of the territory and his own personal debt. He died, a probable suicide, in the fall of 1809.

On request from Clark, who assumed responsibility for the scientific report on the expedition after Lewis' death, Pursh returned some specimens and pencil sketches to the American Philosophical Society; however, he took some expedition specimens with him when he left Philadelphia, first for a position as gardener at the Elgin Botanic Garden in New York City, then to the West Indies, and finally back to Europe. Working in England under the patronage of botanist Aylmer B. Lambert, a founding fellow of London's Linnean Society, Pursh wrote his two-volume *Flora Americae Septentrionalis* (1814), the most comprehensive treatment of American plants to date. In it Pursh described 71 new plants based on plants collected by Lewis and Clark, including the genera *Lewisia* and *Clarkia*.

After completing the flora, Pursh left England for Canada, leaving the Lewis and Clark specimens behind with Lambert. He died four years later in Montreal. Pursh is either villain or hero in the story of Lewis and Clark's specimens, depending on perspective. Apparently he was argumentative, devious, and (more often than not) drunk. Pursh has been much criticized for taking the specimens to England without permission, but had he not done so, they would have languished in obscurity far longer, or perhaps might have been lost entirely.

Lambert's private herbarium was put up for auction after his death in 1842. American botanist Edward Tuckerman attended the auction and purchased half of Lambert's American specimens, including Lewis and Clark's. A specialist in lichens, Tuckerman was not interested in keeping the flowering plant specimens, and he returned these to the Academy of Natural Sciences in Philadelphia in 1856. For a time thereafter, the 47 Lewis and Clark specimens sent by Tuckerman were thought to be the only ones in existence. The others—including those from the first shipment that had gone to Barton and those returned by Pursh to Philadelphia before he left the United States—were lost until 1896, when some were discovered in storage at the American Philosophical Society. The specimens were still bundled in their original pressing papers, and a beetle infestation had damaged some of them.

Despite their history of neglect, today the Lewis and Clark specimens are preserved in a manner befitting their national importance. All 222 specimens have been digitized and are stored in a climate-controlled room designed specifically to hold them. Of this number, the majority is from the western side of the Continental

Lewisia rediviva, from
Curtis's Botanical Magazine
(1863), vol. 89, plate 5395.

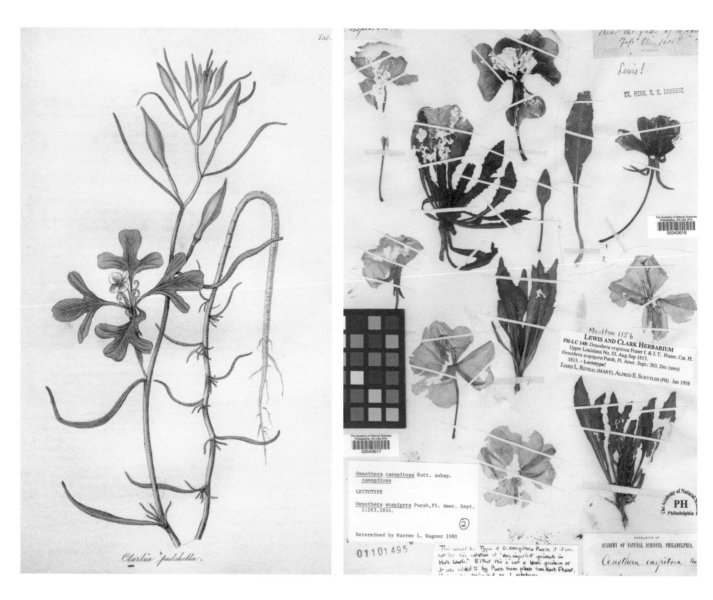

Tab.

Lewis!

EX. HERB. A. B. LAMBERT

Moulton 1156
LEWIS AND CLARK HERBARIUM
PH-LC 148: *Oenothera cespitosa* Fraser f. & J. T. Fraser, Cat. Pl.
Upper Louisiana No. 53, Aug-Sep 1813.
Oenothera scapigera Pursh, Fl. Amer. Sept.; 263. Dec (zero)
1813. – Lectotype!
JAMES L. REVEAL (MARY), ALFRED E. SCHUYLER (PH) Jun 1998

Oenothera caespitosa Nutt. subsp.
     caespitosa

LECTOTYPE

Oenothera scapigera Pursh,Fl. Amer. Sept.
     1:263.1814.

Determined by Warren L. Wagner 1980

01101495

Clarkia pulchella.

PH
Philadelphia

HERBARIUM OF
ACADEMY OF NATURAL SCIENCES, PHILADELPHIA.

Oenothera caespitosa

▲ *Clarkia pulchella*, from Pursh's *Flora Americae Septentrionalis* (1814), vol. 1, plate 11.

◄ Lewis and Clark specimen of *Oenothera cespitosa* (tufted evening primrose). Note the multiple annotations on the sheet: this is one of the specimens that Pursh took to England and Tuckerman purchased at auction.

Divide, with nearly a quarter from what is now Idaho. About 30 specimens, from the first leg of the journey, have never been found.

The story of the Lewis and Clark herbarium specimens follows the same curious trajectory of those of early European explorations—the specimens were gathered and preserved with great effort through the course of a difficult journey, only to be forgotten and neglected later, due to the weakness of the scientific infrastructure at the time. Since their rediscovery and rehabilitation, though, the Lewis and Clark specimens have been well studied by subsequent botanists. Among them are 76 type specimens, cornerstones of our knowledge of the flora of interior North America west of the Mississippi. Beyond that, they are relics of early success of a newly independent nation.

Thomas Nuttall

Some of the plant biodiversity lost to science through the misadventures of Lewis and Clark's specimens was documented a few years later through the efforts of Thomas Nuttall (1786–1859). Born in Yorkshire, Nuttall came to Philadelphia in 1808 to pursue his passion for natural history. Barton, who served as his tutor in plant collecting and identification, eventually sponsored Nuttall's first major collecting trip to Michigan and Canada. On another expedition, Nuttall joined a party of John Jacob Astor's fur trading company that followed a route along the upper Missouri River, similar to that of Lewis and Clark, presenting him with the opportunity to collect in areas from which their specimens were gathered and then lost. A few years later Nuttall explored along the Ohio and Mississippi Rivers through Arkansas and Oklahoma, making plant and animal collections and also detailed descriptions of the native people he encountered. He described his adventures on that trip in his book, *A Journal of Travels into the Arkansa Territory, During the Year 1819.*

▲ Thomas Nuttall during his Harvard years by J. Whitfield.

▶ *Eremogone congesta*, a member of the carnation family, collected by Nuttall (as *Arenaria congesta*) in the Wind River Mountains in 1812. This specimen is from the herbarium of John Torrey, who often mounted different collections of the same species on one sheet.

From 1822 to 1834, Nuttall worked at Harvard University as lecturer and curator of the Harvard Botanical Garden; while there, he produced a textbook on botany and two volumes on birds. But he became restless and left to travel the West Coast, collecting in Oregon and California, and then Hawaii. He eventually returned to England to assume ownership of an inherited property. His major botanical work, *The North American Sylva*, was published in three volumes between 1842 and 1852. Nuttall's goal was not only to build upon Michaux's earlier efforts but also to correct errors in Pursh's *Flora Americae Septentrionalis*. He had an old score to settle with Pursh. When the two met in England in 1812, Nuttall showed Pursh a plant he intended to describe for Barton, who had been a mentor and patron for both men. Shortly afterward, a description of a new genus and species hailing from dry regions of the western United States appeared in print under Pursh's name but based on Nuttall's collection! Nuttall nursed a lingering resentment over having been preempted in publishing the name of this beautiful night-blooming plant, now known as *Mentzelia decapetala*.

Nuttall's peripatetic lifestyle did not lend itself to building a large herbarium of his own. Most specimens from his early trips are in herbaria maintained at the Academy of Natural Sciences in Philadelphia; the rest are widely distributed, with significant numbers in the herbarium of the Museum of Natural History and the Gray Herbarium at Harvard University. Nuttall's specimens are the basis of more than 2,000 new species, including *Cornus nuttallii*, a popular ornamental tree.

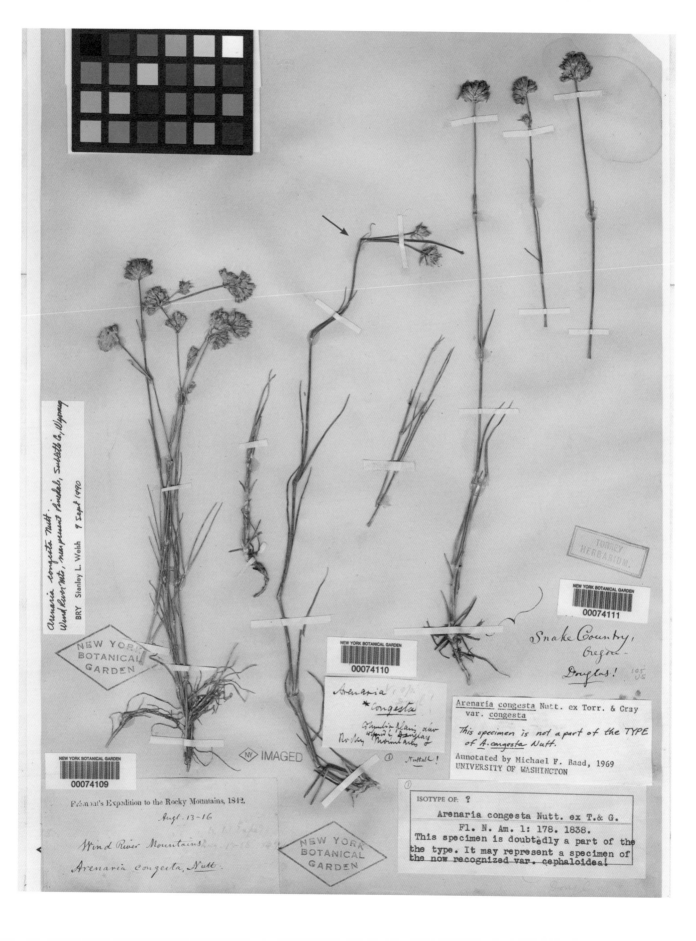

Arenaria congesta Nutt.
Wind River Mtn., near present Pinedale, Sublette Co., Wyoming

BRY   Stanley L. Welsh   9 Sept 1990

TORREY HERBARIUM.

Snake Country,
Oregon.
Douglas!   105
US

Arenaria
* *congesta*

Rocky Mountains

Nuttall!

Arenaria congesta Nutt. ex Torr. & Gray
var. congesta

This specimen is not a part of the TYPE
of *A. congesta* Nutt.

Annotated by Michael F. Baad, 1969
UNIVERSITY OF WASHINGTON

ISOTYPE OF: ?

Arenaria congesta Nutt. ex T.& G.
Fl. N. Am. 1: 178. 1838.
This specimen is doubtedly a part of the
the type. It may represent a specimen of
the now recognized var. cephaloidea!

Frémont's Expedition to the Rocky Mountains, 1842.

Aug. 13-16

Wind River Mountains

Arenaria congesta, Nutt.

Mentzelia decapetala (ten-petal blazingstar), originally published as *M. ornata* in *Revue horticole* (1878), series 4, vol. 50.

Cornus nuttallii (Pacific dogwood), from *Curtis's Botanical Magazine* (1910), vol. 136, plate 8311.

M.S.del.J.N.Fitch lith.

Vincent Brooks Day & Son Lt?imp

New York City: John Torrey's Vantage Point

Primarily a center of commerce in the early years of the nation, New York City was slower to develop a scientific infrastructure than Philadelphia. The first botanical institution there was the Elgin Botanic Garden, founded in 1801 by David Hosack on the site currently occupied by Rockefeller Center. Hosack was both a practitioner and a teacher of medicine, and as family physician to Alexander Hamilton, he was present at Hamilton's duel with Aaron Burr in 1804.

The primary purpose of Elgin, the first public botanic garden in America, was to provide medical students with a source of plants for study, just as Luca Ghini's original garden at Pisa had done. But with Hosack's love of plants, especially exotic species, Elgin's scope extended far beyond what was essential for medical training, resulting in an astounding variety of plants. Hosack eventually ran out of money to support the garden, and in 1810 he sold it at a loss to the state of New York. Elgin was poorly managed thereafter and was eventually abandoned. It is unknown

▲ Painting of John Torrey by Daniel Huntington.

◀ Painting of the Elgin Botanic Garden by an unknown artist, c.1810.

whether or not it had an herbarium; there are specimens at the New York Botanical Garden labeled "Elgin Garden," in what appears to be Hosack's handwriting, but in some cases, dates on the specimens indicate they were collected after the garden closed—perhaps these were collected by Hosack in an attempt to document his failed project.

Although Hosack's garden did not persist, he did set the scene for development of New York City as a botanical center. He was an influence in the early career of John Torrey (1796–1873), a New York City native who became the first American to be internationally recognized as an authority on North American plants. Torrey trained as a physician but did not enjoy medical practice and so developed expertise in chemistry and geology, holding teaching positions in these subjects at West Point Academy, Princeton, Columbia, and New York University. Study of the North American flora was his passion, though. Based largely on his own collections, he compiled a catalog of the plants growing in the vicinity of the city that was published in 1819. A few years later, through correspondence and specimen exchange with Elliott, Muhlenberg, Schweinitz, and many others, young Torrey was able to broaden this study to a flora of the northeastern United States, published in 1824.

Torrey dreamed of writing a flora of the entire United States, and for that reason he agreed to identify the plants resulting from government expeditions to explore the western portions of the country. Mostly run as military operations, these expeditions surveyed the lands west of the Mississippi River to pinpoint national boundaries and determine routes for railway lines. Most also included a scientific team to document the plants and animals encountered. Torrey's former student, physician and geologist Edwin James, served on such a team in the 1819–20 expedition led by Stephen Harriman Long; James was the first person to climb Pike's Peak (named after Zebulon Pike, who first glimpsed the peak in 1806). Torrey described more than 100 species as new to science from James' collections, including the first alpine plants collected in the United States. After that, Torrey received specimens from at least 15 additional western explorations.

From the 1840s on, Torrey gave regular, informal lectures on botany to an audience of interested citizens that grew over time and eventually became the Torrey Botanical Club, now the Torrey Botanical Society. He amassed an herbarium of 65,000 specimens, most of which he prepared himself and annotated with observations and comparisons of species. In 1860, when the collection outgrew the available

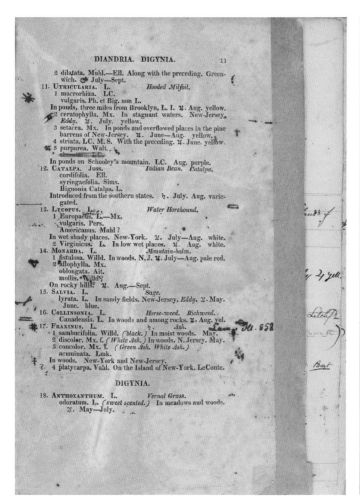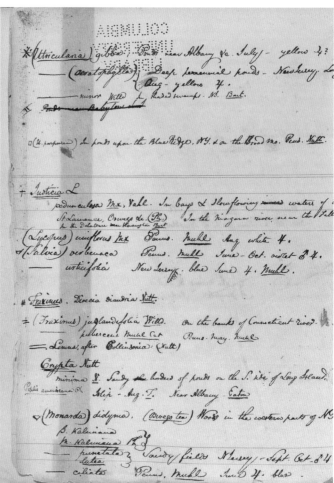

space in his living quarters, Torrey donated his herbarium and library to Columbia. In return the university gave Torrey a workspace in the herbarium as well as a small salary as its curator.

John Torrey died on 10 March 1873. His herbarium and library remained at Columbia until 1898, when they were transferred to the newly founded New York Botanical Garden. Because the specimens sent to him from the many government expeditions to explore the U.S. West were government property, he had returned a set of approximately 20,000 of them to Washington in 1868; at first these were held by the U.S. Department of Agriculture, but eventually they were incorporated into the U.S. National Herbarium at the Smithsonian. The specimens he returned were split from the original gatherings—he maintained a set for his own herbarium as well. Like Schweinitz and Muhlenberg, Torrey was an active correspondent and exchanger of specimens, and his herbarium contains collections made by nearly all early American plant collectors, with the notable exception of Lewis and Clark.

▲ John Torrey's annotated copy of *A Catalogue of Plants, Growing Spontaneously Within Thirty Miles of the City of New-York.*

▶ Torrey herbarium specimen of *Ribes aureum* var. *villosum* (golden currant), collected (as *R. aureum*) by Edwin James on the Long Expedition.

odoratum Wendl.

Ribes aureum, Pursh fl. i. p. 164

Two varieties - one with brownish, the
other with black fruit.

Colo.   Rocky mountains

Dr. James.

Boston: Asa Gray and Harvard University

The herbarium at Harvard was not established until well after Nuttall's tenure there. Its first curator was Asa Gray (1810–1888), student of John Torrey, who became the most influential American botanist of the 19th century. Gray was born in Oneida County, New York. A gifted student, he completed a medical degree before the age of 21 but, having become captivated by plants, never actually practiced medicine. He began a correspondence with Torrey in 1831, and eventually he moved to New York to become Torrey's assistant. The two men developed a lifelong personal and professional friendship. Gray was first hired as a professor at the newly created University of Michigan and was sent by the university on an extended trip to Europe to buy books for the library. Gray also used the opportunity to meet European botanists and study American specimens held in European herbaria. But before he could settle into teaching at Ann Arbor, Harvard University offered him essentially the same position that Nuttall had left some years before, and he took that post instead.

Asa Gray at work, 1868.

Gray was a frequent correspondent of Charles Darwin, and he was an early proponent of the theory of natural selection, though he believed evolution to be guided by a creator. Gray described hundreds of species of plants, most of which were collected by others—like Torrey, he was often described as a "cabinet" botanist, that is, one who works mostly from dried herbarium specimens, not living plants. Gray's *A Manual of the Botany of the Northern United States*, first published in 1848, was a comprehensive and influential guide to the plants of northern North America; the *Manual* has gone through eight editions, the last by Harvard botanist M. L. Fernald revised in 1970. During his lifetime Gray's herbarium grew to be one of the most important collections in the country; he gifted it to Harvard in 1864, and it remains the centerpiece of the Harvard University Herbaria, which today, with an estimated 5 million specimens across six collections, is the largest university-based collection in the world.

Among the other herbaria at Harvard is the Farlow Herbarium, housing specimens of algae, bryophytes, and fungi; it is named for William Farlow (1844–1919), a pioneer in the study of marine algae. Another is the herbarium of the New England Botanical Club, which holds specimens from the many independent collectors in the

Botanist Sereno Watson in Harvard's Gray Herbarium, 1892.

region, including Catherine (Kate) Furbish (1834–1931), the first person to document the flora of Maine. The Oakes Ames Orchid Herbarium, the Economic Botany Herbarium, and the herbarium of the Arnold Arboretum complete the collected Harvard University Herbaria. In addition to these six collections, the Harvard Herbarium Library holds the herbarium of Henry David Thoreau, which consists of approximately 900 specimens collected mostly near his home in Concord, Massachusetts, as well as comprehensive collections of botanical books and archives.

Constantine Rafinesque: An Element of Chaos

Transylvania Seminary (now Transylvania University) in Lexington, Kentucky, was the pioneer institution of higher learning west of the Alleghenies, established in 1780. The first botanist there was Charles Short, a physician and teacher of *materia medica*.

Short had a lifelong interest in plants, and between 1833 and 1837 he published several versions of a catalog of the plants of Kentucky. Asa Gray described Short's herbarium of 15,000 specimens as "a model of taste and neatness" and along with John Torrey named the genus *Shortia* for him, based on *S. galacifolia*, a small white-flowered shrub of higher elevations in the Appalachian Mountains. Short's herbarium was deposited at Philadelphia's Academy of Natural Sciences after his death. Short was joined at Transylvania by one of the most colorful and disruptive figures in American botany, Constantine Samuel Rafinesque-Schmaltz (1783–1840), who was hired as a professor there in 1819. Rafinesque, in sharp contrast to Short, was a model of slapdash and excess.

The son of a French merchant and his German wife, Rafinesque was born in Constantinople. He traveled across the Atlantic when he was 19, arriving in Philadelphia in 1802. There he developed a passion for botany and before long was an acknowledged expert in the field, exchanging information and specimens with all the prominent botanists in eastern North America. His interests expanded to include zoology and anthropology as well.

Rafinesque pursued his natural history studies with an energy bordering on mania. He published more than 1,000 titles during his lifetime, in which he described an astounding 2,700 genera and 6,700 species of plants. Today fewer than 300 of his names are still accepted. Rafinesque had a very narrow species concept, meaning that he viewed even slight variations in such features as leaf size or shape or flower color among otherwise identical plants to be sufficient reason for creating a separate species. This made him a "splitter" in the slang of taxonomists, perhaps the ultimate one.

Many of his contemporaries, though recognizing Rafinesque's intelligence and breadth of knowledge, believed his excessive generation of new plant names derived from lack of careful scholarship and an overabundance of self-promotion. However, unlike many of his contemporaries, who believed that species and genera were fixed entities placed on earth by God, in Rafinesque's view, life on earth represented a series of gradual variations of shapes, sizes, colors, and functions, and these variations become fixed through inheritance. His views were perhaps informed by the writings of 18th-century biologist Michel Adanson, and he also hints at an evolutionary mechanism later described in Darwin's *On the Origin of Species*.

▲ Constantine Samuel Rafinesque-Schmaltz, after the frontispiece illustration from his *Analyse de la Nature* (1815), published in Palermo.

▶ A specimen of *Shortia galacifolia* (Oconee bells), a rare endemic of the southern Appalachian Mountains.

Shortia

Short Metre

Shortia galacifolia, Torr. & Gray.

Hab. Hillside, McDowell County,

Coll. M. E. Hyams, Ap. 6, 1879. North Carolina

In addition to having a narrow species concept, Rafinesque was a careless worker, often renaming a species he had described previously. He acknowledged this but was unapologetic, writing: "I don't know if the Science suffers more from hurry than delay; if two names are given nearly at the same time, we have the choice of the best, if none are given, our knowledge lays buried like the gold of a miser."

Rafinesque's obsession with describing new species made him a challenging companion, as demonstrated by a story told by John James Audubon. Rafinesque spent much of the year 1818 collecting plants and animals along the Ohio River, often visiting fellow naturalists along the way. He appeared at Audubon's home in Henderson, Kentucky, without warning, wishing to discuss what he believed were new species of fish he had found in the region. Audubon described him as disheveled and wearing clothes filthy from travel and fieldwork, but he invited him to stay nonetheless. Rafinesque declined the opportunity to bathe or put on a change of clothes provided by his host. During the night, Audubon was awakened by a noise coming from Rafinesque's room. In his *Ornithological Biography* (1832), Audubon described the scene he encountered when he went to investigate:

> [T]o my astonishment I saw my guest running about the room naked, holding the handle of my favourite violin, the body of which he had battered to pieces against the walls in attempting to kill the bats which had entered by the open window . . . . I stood amazed, but he continued jumping and running round and round, until he was fairly exhausted, when he begged me to procure one of the animals for him, as he felt convinced they belonged to "a new species." Although I was convinced of the contrary, I took up the bow of my demolished Cremona, and administering a sharp tap to each of the bats as it came up, soon got specimens enough.

It should be noted that apparently Audubon was known to exaggerate for the purpose of a good story, and despite the incident, Audubon described Rafinesque as "a most agreeable and intelligent companion." Audubon may have harbored some resentment over the violin, however, because afterward he began to send Rafinesque descriptions and sketches of fictitious animals of his own imagination. Rafinesque described about 30 of these based on Audubon's notes, some of them more than once!

Rafinesque's teaching career at Transylvania University ended in 1826 following a dispute with the president, and afterward he returned to Philadelphia with 40 crates containing his herbarium of some 35,000 to 50,000 specimens. For the rest of his life he supported himself chiefly by selling specimens and books. He supplemented this income with his patented "divitial invention," a banking system of issuing stocks

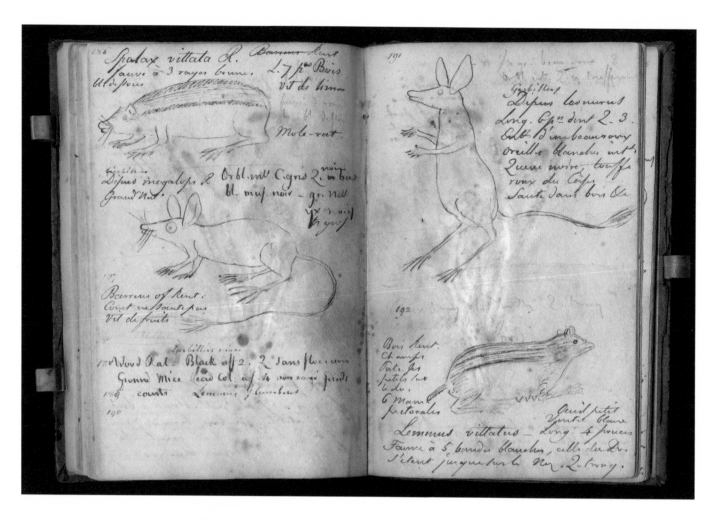

Rafinesque's notes on Audubon's fictitious species.

and bonds in divisible, easily traded units, and through the sale of Pulmel, an herbal medicine he'd invented for tuberculosis. The medicine was based on herbs he purchased from the Shakers, but the formula was kept secret. Rafinesque continued to publish voluminously, but most of his writings at this time were self-published; the sheer volume and questionable quality of his work had caused most of the scholarly publishers to refuse it.

Rafinesque died in Philadelphia of stomach cancer in 1840 and was buried in a cemetery created for travelers or others who for some reason could not be buried in a church cemetery. His herbarium was offered for sale after his death for $6,000. There were no takers, so his specimens were stored in a rented attic. Eventually Philadelphia pharmacist and botanist Elias Durand purchased them for $45. Durand described the state of herbarium in a letter to Charles Short:

[I]t was truly a dunghill, from the circumstance that it had remained three years in a stable garret, where a numerous colony of rats found it a convenient hiding and [breeding] place. It required indeed a botanical zeal equal to mine to consent [to] transfering to a decent house 6 cart-loads of that prodigious mass of corruption. Happily it was not a season of yellow fever! Not being fit to place in a room, I removed it to an obscure cellar, where, for a whole winter, I was occupied, by candle light, to separate the good from the rotten. . . . Rafinesque was the worst collector I ever knew.

Only 275 specimens survived Durand's culling process, meaning that many of Rafinesque's type specimens, those upon which his new species were based, were discarded. Durand donated the remaining Rafinesque specimens to Philadelphia's Academy of Natural Sciences, although there are some at both Harvard's Gray Herbarium and the New York Botanical Garden's Steere Herbarium. His specimens are sometimes ample, sometimes very sparse.

Rafinesque's excesses as well as Durand's unfortunate curatorial practices have caused many headaches for later workers. Before describing a plant as new species, good scholarship dictates that a botanist must review all pertinent literature to be sure the plant has not already been named. The vast number of Rafinesque's names and obscure publications, compounded by the absence of type specimens or detailed descriptions, created serious problems for those interested in documenting North American plant diversity. To help overcome the obstacle of Rafinesque's legacy of plant names, botanist Elmer Merrill of Harvard University took on the monumental task of compiling Rafinesque's publications and the plant species he described. In the introduction to this work, published in 1949, Merrill wrote: "[M]y frank conclusion is that in taxonomy and nomenclature we would have been infinitely better off today had Rafinesque never written or published anything appertaining to the subject."

While his reputation among biologists may have been poor, Transylvania University has maintained a fascination with Rafinesque. When the cemetery where he was interred in Philadelphia was to be destroyed 80 years after his death, Transylvania University arranged to have his remains brought to Lexington for entombment under the steps of Old Morrison, one of the original campus buildings. Unaware of the cemetery's practice of burying multiple bodies in one grave, the university reclaimed the wrong body, and the remains in Rafinesque's tomb actually belong to someone else. However, this mix up apparently has not diminished the symbolic importance of the monument to the university. Each year, students there celebrate "Rafinesque Week" leading up to Halloween. The event purports to stave off the curse of Rafinesque—legend has it that Rafinesque placed a curse on the university after the president searched his rooms without his consent. The following year, the

TRANSYLVANIA UNIVERSITY (MORRISON COLLEGE), LEXINGTON.
(Kentucky University, since 1866.)

▲ Rafinesque specimen obtained by John Torrey, labeled "*Gentiana trachitonia* var. *biflora*," a name that was never published.

◀ A view of Transylvania University's iconic Old Morrison building (and its steps), from *Collins' Historical Sketches of Kentucky: History of Kentucky* (1874).

college burned, and the president died of yellow fever, providing evidence of the power of the curse, to some. Among the activities of Raf Week is an auction for the chance to spend the night in Rafinesque's tomb!

St. Louis: Henry Shaw and George Engelmann

The convergence of two European immigrants to the St. Louis area led to the establishment of the most important center of botanical study in the U.S. Midwest—the Missouri Botanical Garden. One of these was Englishman Henry Shaw (1800–1889) who emigrated in 1819 and proceeded to build a fortune in hardware and cutlery that later expanded to include agriculture, mining, real estate, and furs. On the outskirts of the city, he purchased land that he cultivated using slave labor, a practice he abandoned before the Civil War. In an act of philanthropy unusual at the time, he decided to use some of his land for a public garden.

Development of Herbaria in the United States | 133

George Engelmann (1809–1884), a German physician with a lifelong interest in botany, settled in St. Louis in the 1830s. He had a particular interest in the cactus and grape families and published scholarly articles on both groups. His expertise in grapes played a role in saving the French wine industry in the 1870s. At this time, French vineyards were under attack from an aphid-like insect in the genus *Phylloxera*. Working as a consultant to the French government, Engelmann was able to find a suitable North American grape species, *Vitis riparia*, from the Mississippi River Valley, which was resistant to the pest. He arranged for the shipment of this species to France, where it proved very successful as a rootstock for French grapes and is still used as such today.

Engelmann built a personal herbarium of nearly 100,000 specimens through his own collecting and exchange with fellow botanists. He also purchased an important European herbarium, that of Johann Jacob Bernhardi (1774–1850), a German physician whose collection of nearly 60,000 specimens held a worldwide representation of species. As a leader of the scientific community in St. Louis and also editor of a well-regarded German newspaper, Engelmann was in a position to persuade Henry Shaw to include a scientific component in his public garden. Shaw's Garden (eventually renamed the Missouri Botanical Garden) opened in 1859, and the Engelmann and Bernhardi herbaria became the nucleus of what is today one of the largest plant collections in the United States (and the world), and one that is strongly influential in plant science and horticulture.

▲ "View along the Gila (*Cereus giganteus*)" (now *Carnegiea gigantea*) by Paulus Roetter, from George Engelmann's *Cactaceae of the Boundary* (1859).

▶ Specimen of the freshwater alga *Chara zeylanica*, collected by Engelmann "in the Meramec River, below St. Louis" in 1842.

ANNOTATION

<u>Chara zeylanica</u> Klein ex Wild.

DET. R. D. WOOD _____ 19 60
UNIVERSITY OF RHODE ISLAND

ANNOTATION

Chara zeylanica var. Michauxii

DET. R. D. WOOD *Rd* *mar. 14* 19 55
UNIVERSITY OF RHODE ISLAND

Chara polyphylla
var Michauxii AB

In the Maramee River, below
St Louis,

George Engelmann, M. D.
ST. LOUIS, Mo. *Sept 1842*

No. 1. Residence Henry Shaw.	No. 4. Museum.	No. 7, 8 and 9. Summer Houses.	No. 12. Gardner's House.
No. 2. Gardener's Rooms.	No. 5. Pavilion.	No. 10. Mitchel's Winter Garden.	No. 13. Park Superintendent's House.
No. 3. Mausoleum.	No. 6. Palm House.	No. 11. Casino.	No. 14. Entrance Lodge to Garden.

SHAW'S GARDEN.

▲ Shaw's property on the outskirts of St. Louis. He described the land as "uncultivated, without trees or fences, but covered with tall luxuriant grass, undulated by the gentle breeze of spring." Map by Camille & Dry, 1875.

◄ The Missouri Botanical Garden's Bayer Center, which houses part of the herbarium.

Pioneers of California Herbaria: Brandegee, Eastwood, and Jepson

Archibald Menzies, while serving as surgeon on two English exploring expeditions, collected in Vancouver Island, southern Alaska, Monterey, and San Francisco between 1783 and 1795; most of his specimens are held at the Sloane Herbarium and the Royal Botanic Gardens, Kew. Around the same time, Thaddeus Haenke, a naturalist on the Malaspina expedition (1789–94), collected in the vicinity of Monterey Bay. Czech botanist Karl B. Presl described the new species collected on this expedition in *Reliquiae Haenkeanae* (1825–35). Haenke's specimens are maintained in the herbarium of the National Museum in Prague.

It was not until the Gold Rush transformed the quiet mission town of Yerba Buena into the city of San Francisco that a center for botanical study developed on the West Coast of the United States. The California Academy of Sciences was founded in 1853 with the mission of making "a thorough systematic survey of every portion of the State and the collection of a cabinet of her rare and rich productions." The University of California, established in 1868, also had the goal of documenting the natural history of the state. The founding documents included the provision for a "museum of the university" intended to house collections made by the California Geological Survey; the survey, instituted to find new natural resources in the state as the Gold Rush wound down, was charged with documenting and preserving samples of all minerals, plants, and animals.

The eastern botanical establishment led by John Torrey and Asa Gray dominated the study of the American flora in the mid-1800s; however, with the establishment of the California Academy of Sciences and the University of California, a cohort of West Coast botanists began to challenge the eastern authority. The westerners' work was often dismissed by East Coast botanists, not unlike the way European botanists had dismissed the early work of American naturalists. Gray described Albert Kellogg, founding member and first botanist at the California Academy of Sciences as "a good meaning soul" but "a nuisance in the science." Regardless of the quality of his work, Kellogg was responsible for an addition to the Academy's charter indicating that the institution "highly approved the aid of females in every department and earnestly invited their cooperation." The California Academy of Sciences thereby became one of the first institutions in the country and possibly the world to recognize and encourage women scientists, a decision that paid off for the institution in spades, due to the efforts of two highly accomplished and energetic women that it employed.

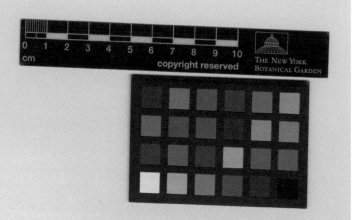

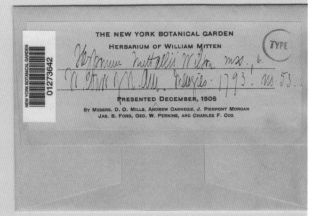

Hypnum nuttallii Wilson TYPE
Bryol. Brit. 334, 339, 1855

▶ The California Geological Survey team, December 1863 (from left): Chester Averill, assistant; William M. Gabb, paleontologist; William Ashburner, field assistant; Josiah D. Whitney, state geologist; Charles F. Hoffmann, topographer; Clarence King, geologist; and William H. Brewer, botanist.

◀ A moss collected by Menzies on the "West Coast of N. America," 1793. This specimen was likely sent by Hooker to William Mitten, whose herbarium was later purchased by the New York Botanical Garden. It may be a type duplicate (isotype) of *Hypnum nuttallii* (now *Homalothecium nuttallii*).

The first was Mary Katharine Brandegee (1844–1920). She was born in Tennessee but moved with her family first to Salt Lake City and then to the Sierra Nevada foothills in California. After a shortlived marriage to constable Hugh Curran, she pursued her dream of becoming a medical doctor; she enrolled in the University of California, the third woman to matriculate there, obtaining her medical degree in 1878. Following a pattern so common among other physicians-turned-botanists, she became fascinated with plants while studying their medicinal properties, so instead of opening a medical practice, she joined the staff of the California Academy of Sciences, becoming curator of the herbarium in 1883. She quickly gained a deep mastery of California plants through her extensive collecting in California and Nevada, often traveling by train as far as possible and then walking 20 miles or more per day to obtain specimens. She fell "insanely in love" with fellow botanist Townshend Stith Brandegee when he visited the Academy in 1886, and together they continued the life path both had chosen individually, starting this journey in earnest by walking 500 miles from San Diego to San Francisco on their honeymoon in 1889, collecting plants along the way!

Kate was imposing physically, as well as forthright and outspoken, yet scientifically at least, she was cautious. She was not quick to describe new species, and she was more inclined to defer to the eastern botanical authority than colleagues such as Edward L. Greene, her neighbor across the bay at UC Berkeley. In the journal

Zoe that she and Townshend helped to found, she offered a spirited defense of Gray's work in response to a scathing commentary by Greene.

However, in an exchange with botanist Nathaniel Britton of Columbia University, Kate demonstrated that her feminine and western pride could be ruffled by a perceived slight from an easterner. She felt that Britton either ignored or downplayed the importance of a number of western publications in a catalog he published of American floristic works in 1889. Kate took Britton to task for these omissions in a caustic review in *Zoe*. Regrettably, Britton opened his response to the review, also published in *Zoe*, with a quote from Shakespeare's *Taming of the Shrew* in which Petruchio describes just how unimportant a woman's opinion is to a man as great as he; he went on, however, to offer only a vague defense of his omissions in the article. In a rejoinder to Britton's response, printed immediately following it, Brandegee chided him for his sexism, ending her short but sarcastic comments with: "We are under the impression that among gentlemen the *argumentum ad hominem* is considered to be extremely bad form; the *argumentum ad mulierem* is however apparently admissible."

▲ Kate and Townshend Brandegee.

▶ Specimen of *Astragalus pycnostachyus* var. *pycnostachyus* (marsh milkvetch), a coastal California endemic, collected by Katharine Brandegee in 1907.

FLORA OF CALIFORNIA

Astragalus pycnostachyus Gray
Bolinas Bay.

Katharine Brandegee June 30, 1907

A substantial inheritance bequeathed to Townshend Brandegee led Kate to give up her salary so that she could hire Alice Eastwood, a talented young botanist from Colorado, to replace her. Katharine and Townshend then retired to San Diego, where they created an herbarium, library, and large native plant garden. From there they continued their botanical explorations, including trips to Baja California. After the 1906 San Francisco earthquake, which largely destroyed the Academy's collections, the Brandegees moved back to the Bay Area, donating their herbarium of more than 76,000 specimens and their library to the University of California, where they worked for the rest of their lives. Together they made some 20,000 collections and described over 1,000 new genera and species.

Alice Eastwood was very much Kate's kindred spirit. Eastwood shared her fascination with plant life and love of strenuous fieldwork. Eastwood was born in Toronto in 1859 but moved to Denver as a young teen. Her early love of nature, especially plants, grew during long hikes in the Rockies. She was a talented student, graduating from high school as class

Alice Eastwood, c.1910.

valedictorian; however, her family's precarious financial situation precluded the possibility of higher education for her. Instead she taught high school but continued to spend as much time as possible exploring the plants of the Rocky Mountains on her own, and she developed a local reputation for her trail experience and plant knowledge. When Alfred Russel Wallace visited Denver in 1887, it was Eastwood who took the famous tropical explorer, then in his sixties, on a three-day hike to Grays Peak, introducing him to Colorado's diverse and beautiful alpine flora near the 14,000-foot summit. Grays Peak and the adjacent Torreys Peak were first ascended and named by botanist Charles Parry, who named them to honor his botanical mentors.

Shrewd investing of her modest teacher's salary in real estate generated enough income that by the time Eastwood was 31 she could quit her teaching job and devote herself full-time to botany. She used her newfound freedom to travel to San Francisco

View of the Continental Divide Trail and Grays Peak from Torreys Peak.

to study specimens at the Academy. The Brandegees immediately invited her to contribute to *Zoe*, and later invited her to join the Academy's staff, at first to help organize the collection and shortly thereafter to replace Kate Brandegee as curator. Eastwood learned the California flora through study of existing specimens as well as extensive fieldwork, on foot and horseback, through the Sierra Nevada and Coast Ranges. She funded her fieldwork herself and also used her own funds to buy collections for the Academy's herbarium.

Through her efforts, the herbarium of the California Academy of Sciences became the most complete record of western plant life, consulted frequently by other botanists. The Brandegees were apparently somewhat negligent curators, but under Eastwood's care, the collection was reorganized for the protection of the specimens and ease of use by researchers. Eastwood started the practice of filing the type specimens separately from the general collection. She rehoused the types in a lightweight metal cabinet that could be lowered out of a window in case of fire.

Eastwood's energy and enthusiasm for the study of plants and her dedication to the Academy's herbarium were well known in California but gained worldwide appreciation through her heroic efforts to save the herbarium during the earthquake and fire in San Francisco on the morning of 18 April 1906. Eastwood felt the tremor that morning, but since it was not strong in her Nob Hill neighborhood, she went to work that morning with no particular anxiety; however, when she reached the

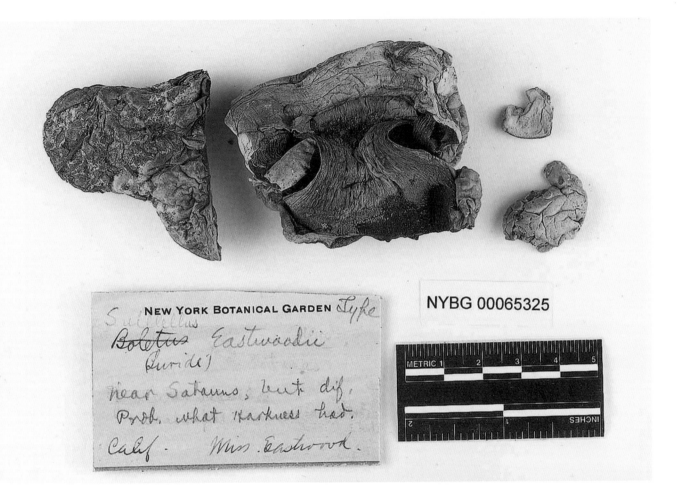

NEW YORK BOTANICAL GARDEN *Type*

NYBG 00065325

Suillellus
Boletus Eastwoodii
(luride)
near Satanus, but dif.
Prob. what Harkness had.
Calif.   Miss. Eastwood.

Academy, then located on Market Street, she saw that the building was damaged, and worse, the surrounding buildings were on fire! She and the few other Academy staff who had also arrived by then sprang into action. The marble stairway leading to the floor where the herbarium was housed had collapsed, but its bronze banister remained intact. She walked up the banister, placing her feet between the rungs to ascend to the sixth floor, where the herbarium was located. From the herbarium windows she could see that fire had reached the adjacent building and that she had only minutes to save what she could. She chose the approximately 1,500 type specimens and her favorite Zeiss lens. The type cabinet was damaged and could not be evacuated as she had planned. So, she bundled the specimens in a large piece of fabric, secured it with rope, and lowered them to ground level. By the time she got back downstairs, police and firefighters were desperately trying to clear the area as the fire closed in. Ignoring their entreaties for her to flee, Eastwood secured the specimens on a horse-drawn cart and had them taken to safety, first to Russian Hill, then, when fire threatened that area, to the Presidio. Later that day her own home burned—she was not able to save anything there. Eastwood gave an account of the incident in a letter to *Science* magazine published on 25 May 1906. In it she wrote: "I do not feel the loss to be mine, but it is a great loss to the scientific world and an irreparable loss to California. My own destroyed work I do not lament, for it was a joy to me while I did it, and I can still have the same joy in starting it again."

After the earthquake, the herbarium at UC Berkeley offered Eastwood work space, and she began, using her own funds, to recollect the plants that were lost and to purchase books to rebuild the Academy's herbarium and library. She also visited eastern herbaria, including the U.S. National Herbarium in Washington, D.C., and Harvard's Gray Herbarium. When the Academy reopened its doors in 1916 at its new quarters in Golden Gate Park, Eastwood went right to work rebuilding the collection. In this new setting, Eastwood could also indulge her interest in teaching the public about plants—she arranged for weekly educational displays of flowers in the Academy's main gallery and taught botany to the public as well as to Golden Gate Park's gardening staff.

For the remainder of her long life, Eastwood continued to collect plants, build the herbarium, and share her knowledge and love of plants. Under her care the new herbarium grew from the nucleus of saved type specimens to more than 300,000 accessions, collected by her and others, by the time of her 80th birthday. Eastwood authored over 300 publications during her career, 200 of these after age 50. At the age of 91 Eastwood traveled to Sweden to the International Botanical Congress, where she was made honorary president of the Congress and was seated in Linnaeus' chair. Eastwood was offered honorary college degrees on several occasions; however, she would not accept these, dismissing the notion that academic degrees need to be associated with significant scientific contributions. Eastwood ignored the expectations in her time for the appearance or comportment of women. When other women remarked to her that she was able to do things they could not, she was known to quote Ralph Waldo Emerson in response: "Always scorn appearances, and you always may." She died at the age of 95, and several years later, her beloved California Academy of Sciences named the Alice Eastwood Hall of Botany, a public exhibit about plants, in her honor.

CLOCKWISE,
FROM TOP LEFT
Pre-earthquake photo of
the California Academy
of Sciences on Market
Street, San Francisco.

View of the Academy
after the earthquake from
the central court of the
main floor.

Alice Eastwood (1859–1953)
with her microscope.

Alice Eastwood saved many
type specimens from the
California Academy of
Sciences fire, including this
one of *Agave sebastiana*.

Willis Linn Jepson, plant press in hand, collecting in the Sierra Nevada, 1911.

Willis Linn Jepson (1867–1946), a California native, was already very knowledgeable about plants when he entered UC Berkeley as an undergraduate. He obtained his Ph.D. in botany there in 1899 and was immediately hired as a professor. His *Manual of the Flora of California* (1925) was for many years the standard identification tool for California plants. In addition to his interest in documenting the plants of California, Jepson was a dedicated conservationist; along with John Muir and others, he was a co-founder of the Sierra Club in 1892. He was also committed to educating the public about California plants. When he bequeathed his extensive herbarium to the university, he stipulated that it remain separate from the rest of the UC Berkeley herbarium, and that it be dedicated to three main goals: updating the *Manual*, completing a flora of California with color illustrations, and maintaining and expanding the herbarium and library. The Jepson Herbarium, housed alongside and co-managed with the UC Berkeley herbarium, has stayed true to the mission envisioned by Jepson. The most recent version of the *Manual*, augmented by the digitized specimen data from the Jepson and other California herbaria, is available online: ucjeps.berkeley.edu/eflora. Several species of the California flora are named for Jepson, including *Ceanothus jepsonii*, which is endemic to the San Francisco Bay area and Coast Ranges to the north.

American Exsiccatae

In the United States as in Europe, the study of non-vascular plants and fungi lagged somewhat behind that of the more conspicuous representatives of the flora, though not through lack of interest—the herbaria of Banister, Clayton, and Michaux all contain bryophytes, as did those of Bartram and Schweinitz. However, the small size of the plants and lack of reference works was (and remains!) an impediment to documenting these organisms.

American students of vascular plants generally shared their incremental discoveries and assessments of their flora through published articles and books. Knowledge about American cryptogams was shared similarly, but specialists in these groups took advantage of another means of documenting their organisms found there: the aforementioned exsiccatae. Authors (or editors, as they were usually called) of exsiccatae found this a desirable means of distributing the results of their research: they didn't need to produce detailed descriptions or drawings of microscopic features, nor did they require the services of a publisher. It was not uncommon for type specimens to be distributed as part of exsiccatae, and before the rules of nomenclature forbade it, a set of exsiccatae was sometimes the place of publication of the new species. If offering sets for sale could make a profit, editors could fund future collecting and specimen preparation. Exsiccatae appealed to purchasers because they could examine and dissect a set of authentically determined specimens in order to learn their features and compare them directly with other collections. Sets of ten specimens were called a "decade," sets of 100 specimens were referred to as a "century." Though some were issued only once or a few times, other exsiccatae continued for many years, some to this very day.

Bound volumes of Collins' *Phycotheca Boreali-Americana*, with one volume opened to show a specimen of red algae.

An important cryptogamic exsiccata issued in the early years of documenting the plants and fungi of North America was *Phycotheca Boreali-Americana* (A Collection of North American Algae) by Frank Shipley Collins (1848–1920). Collins worked for more than 30 years at the Boston Rubber Shoe Company as an efficiency expert and bookkeeper, but he devoted all his leisure time to the study of natural history in his native New England. He came to focus his interests on marine algae through his wife, Anna Holmes Collins, and her friendship with Maria Bray. Bray's husband had been a lighthouse keeper on Thacher Island off of Cape Ann. To pass the time in this lonely outpost, Bray began to collect and prepare specimens of algae, or sea mosses as she called them. She pressed them in attractive arrangements on cards, each labeled with a scientific name, and offered them for sale; some of the cards were purchased by Anna Collins. When Anna showed the cards to Frank he was much impressed with the presentation but knew from his own study that the determinations were incorrect. In an effort to help his wife's friend provide the correct names on her sea moss cards, he dove deeply into the meager literature available on American algae, and, once captivated by these beautiful organisms, devoted every spare moment he could thereafter to studying them. Collins corresponded widely with professional phycologists around the world and made numerous field excursions, including to Bermuda.

Algae are notoriously difficult to prepare as specimens. One must float the sample in a shallow tray of water (seawater, for marine algae), arrange it so the branches are not overlapping, and then insert the mounting paper into the tray and lift the specimen on it, making any final rearrangements in the layout before pressing. The specimen must be dried immediately to prevent discoloration and decay, which requires frequent changes in the blotter paper used between specimens to capture the moisture. Thus, the time and effort necessary to produce *Phycotheca Boreali-Americana*, published between 1895 and 1919, was monumental. In total, 80 sets were produced, each consisting of about 2,300 specimens. Given the paucity of literature and illustrations available for these organisms, *Phycotheca Boreali-Americana* was the most authoritative source of information

A postcard for the Boston Rubber Shoe Company, where Frank Shipley Collins spent his career as bookkeeper.

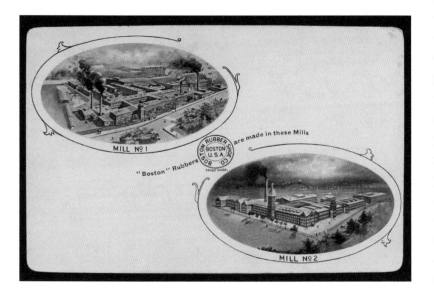

concerning them. It was distributed to 37 herbaria in North America, and for inland institutions, these beautiful specimens provided students with perhaps their only experience of the diverse plant life of the nation's coastlines.

For fungi, the best-known American exsiccatae were those prepared by Job Bicknell Ellis (1829–1905), the first American student of fungal plant diseases. Born in Potsdam, New York, Ellis attended Union College in Schenectady, New York, after which he worked first as a teacher and then joined the Union Army during the Civil War. He and his wife, Arvilla, eventually settled in Newfield, New Jersey, where he lived the rest of his life. Botany had been a serious avocation for Ellis since childhood. When he happened to see a copy of Henry Ravenel's *Fungi Caroliniani*, the first American exsiccata, issued between 1852 and 1860, he was inspired to take up the study of fungi. He began a correspondence with Ravenel, a plantation owner in South Carolina. Ravenel advised him on collecting and preparing fungal specimens and suggested which European literature he should purchase for reference. Following Ravenel's advice, Ellis at first sent many of his specimens to British mycologist Mordecai C. Cooke, who published the new species under shared authorship. Eventually, though, differences in species concepts led Ellis to terminate this arrangement and publish on his own.

Ellis produced three sets of fungal exsiccatae, with Arvilla doing most of the preparation. The first, *Fungi Nova-Caesariensis* (Fungi of New Jersey), consisted of a single century, one of which he gave to William Farlow at Harvard University. Impressed with the effort, Farlow encouraged Ellis to broaden the scope to the whole of North America and provided funding so that Ellis could issue the specimens in bound books instead of loose sheets. Ellis' second set of exsiccatae, *North American Fungi*, was a success—in total, about 200,000 specimens were sold to subscribers, providing enough return to support the Ellis family for a while. For the second half of *North American Fungi*, Ellis joined forces with Benjamin M. Everhart, a wealthy merchant and keen mycologist living nearby in West Chester, Pennsylvania. They

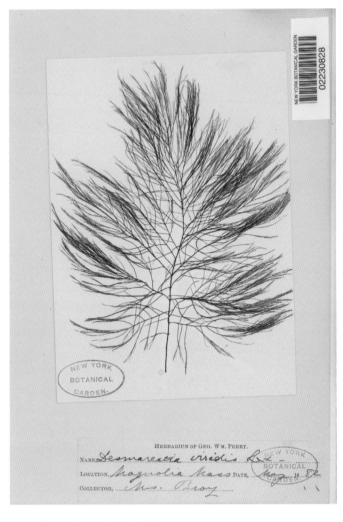

▲ A specimen of the common brown alga *Desmarestia viridis*, collected and beautifully prepared by Maria Bray. It came to the New York Botanical Garden from the herbarium of George William Perry, who was active in the Boston area at the turn of the 20th century

▶ *Eucheuma isoforme*, a red alga collected by Collins in 1912 in Bermuda and distributed as *Phycotheca Boreali-Americana* specimen no. 1886.

Eucheuma isiforme (C.Ag.) J.Ag.
var. isiforme

Donald P. Cheney   7 X 75
University of South Florida, Tampa

Phycotheca Boreali-Americana.   Collins, Holden and Setchell.
                   Algae of Bermuda.

1886. Eucheuma isiforme (Ag.) J. Ag.

J. G. Agardh, Nya Alger Mexico, p. 16, 1847; Sp. Alg., Vol. II,
    p. 627, 1852; Vol. III, p. 600, 1876.
  Harvey, Nereis Bor. Am., part 2, p. 118, Pl. XXIV, 1853.
Farlow, Proc. Amer. Acad., Vol. X, p. 366, 1875; Report of
    U. S. Fish Commission for 1875, p. 697, 1876.
        De Toni, Syll. Alg., Vol. IV, p. 370, 1900.
Sphaerococcus isiformis Agardh, Sp. Alg., Vol. I, part 2, p.
    271, 1822.

    On exposed muddy rocks, Tuckertown, April 25, 1912.
                                    F. S. COLLINS.

    The same species, from Florida, was distributed as P. B.-
A., No. 92.

also collaborated on the third set of exsiccatae Ellis produced, *Fungi Columbiani*, consisting of 60 sets of 14 centuries.

Ellis described some 4,000 species of fungi during his lifetime, and duplicates of many were included in the exsiccatae sets. He did not travel much himself but had a wide correspondence, receiving specimens from throughout North America and beyond. Among the people who sent him specimens was George Washington Carver. Born a slave in Diamond Grove, Missouri, in the 1860s, Carver obtained his Ph.D. from Iowa State University. He is best known for his research into and development of peanuts and other agricultural crops; however, Carver began an interest in fungi while a graduate student at Iowa that was nurtured during his Tuskegee years by Ellis, among others. In 1902, Ellis and Everhart published "New Alabama Fungi," an article that listed 60 species received from Carver, including two new species named for him. The New York Botanical Garden has more

Staff of the agricultural school at Tuskegee Institute (now Tuskegee University), with George Washington Carver front and center, c.1902.

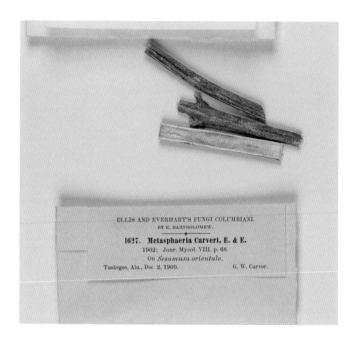

ELLIS AND EVERHART'S FUNGI COLUMBIANI.
BY E. BARTHOLOMEW.

1627.   Metasphaeria Carveri, E. & E.
1902: Jour. Mycol. VIII. p. 68.
On *Sesamum orientale.*
Tuskegee, Ala., Dec 2, 1900.   G. W. Carver.

A fungal specimen collected by Carver in Tuskegee, Alabama. Carver sent the specimen to Ellis, who (with Everhart) distributed it in *Fungi Columbiani* as the type of *Metasphaeria carveri.*

than 200 specimens collected by Carver, most of which were in *Fungi Columbiani*, and thus Carver's collections are distributed widely across North America and Europe.

At an advanced age, Ellis sold his private herbarium of 80,000 specimens and his library to Nathaniel and Elizabeth Britton for their nascent herbarium at the New York Botanical Garden. On 15 March 1896 he wrote to Elizabeth: "The Her. [barium] is all packed and ready to go. A part of my life will go with it, but the thought that it is only temporarily lost to me makes the parting endurable." The comment about the temporary nature of the separation betrays the Ellises' adherence to Spiritualism, a religious movement that was very popular at the time.

Exploration Beyond National Borders

While the exploration of plant life in the American West was still in its early stages, Americans began to collect plants beyond their country's borders as well. The United States Exploring Expedition (1838–42) was an exploring and surveying expedition of the Pacific Ocean, authorized by Andrew Jackson and commanded by Charles Wilkes. It was a lavish undertaking. The expedition's six ships and 342 crew, including nine scientists and artists, launched two years after the completion of the voyage of the *Beagle*. Aspiring to be recognized as a world power on par with European countries, the U.S. government hoped the expedition would increase American prestige by discovering and describing unknown parts of the world and by making scientific collections of previously unknown plants and animals. Among the large crew of the expedition was a scientific team that included naturalists, a minerologist, and a philologist. The botanists of the expedition were William Brackenridge and William Rich. Asa Gray was first asked to be the botanist, but having just been offered an academic position at the newly founded University of Michigan, he declined the post.

The expedition visited Madeira, Rio de Janeiro, Tierra del Fuego, coastal Chile, and Peru. It then sailed to Antarctica, surveying 1,500 miles of coastline of what is still known as Wilkes Land, before heading north to Australia, Samoa, Fiji, and Hawaii. Returning to North America, it explored coastal and inland areas of

Washington, Oregon, and California. The expedition did not run entirely smoothly: one ship was lost with all hands, and hostilities with indigenous people in Fiji led to the loss of several sailors and 80 Fijians. Wilkes was apparently a harsh commander who had poor rapport with his subordinates; he was court-martialed upon his return and charged with illegal punishment of his crew.

But the expedition was a success scientifically, yielding approximately 50,000 specimens of 10,000 species. The government had strict rules about the specimens: everything belonged to the government and nothing could be published without permission. However, there was little incentive for expedition members to identify and prepare the specimens once the expedition was completed, and the material languished and in some cases was misused—some specimens, such as seedlings of Norfolk Island pines grown from seeds collected on the expedition, were given away as gifts to politicians. Identification of the expedition's plant specimens fell to Torrey and Gray, as well as botanist William Darlington of West Chester, Pennsylvania; Torrey's report concerning the plants collected on North America's Pacific Coast was published posthumously in 1874.

Collections from the Wilkes Expedition (as it is also known) eventually found a home at the Smithsonian Institution. Founded in 1846 with a bequest from British scientist James Smithson, the Smithsonian was intended to be a center for scientific research as well as a depository for government collections and others made in the Washington, D.C., area. The botanical collections from the Wilkes and other government expeditions are stored at the U.S. National Herbarium there.

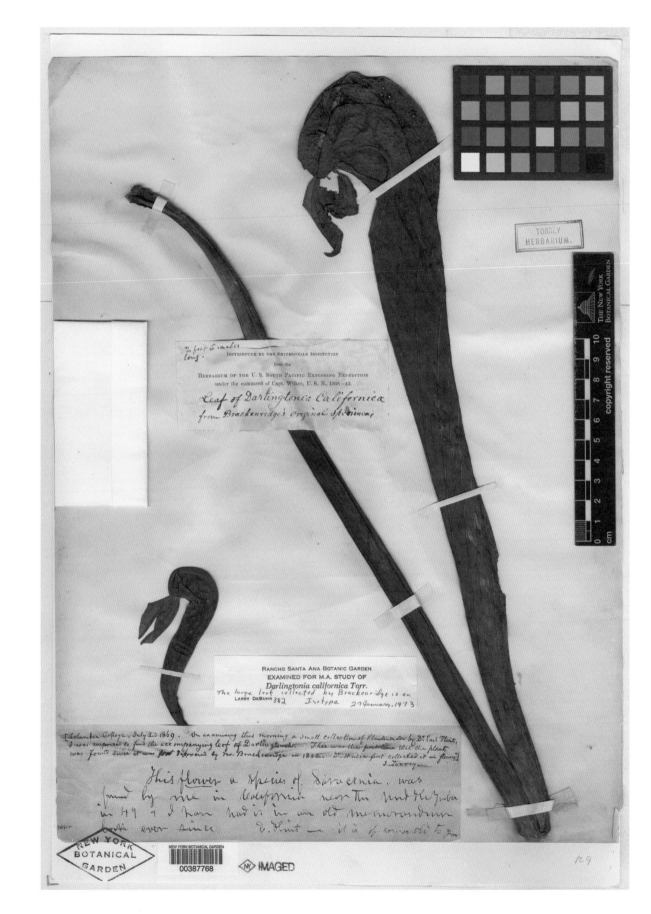

2 feet 6 inches
long.

Leaf of Darlingtonia Californica
from Brackenridge's original specimens

Rancho Santa Ana Botanic Garden
EXAMINED FOR M.A. STUDY OF
Darlingtonia californica Torr.
The large leaf collected by Brackenridge is an
LARRY DeBUHR 882   Isotype   27 January, 1973

Columbia College, July 20 1869. On examining this morning a small collection of Plants made by Dr Carl Flint,
I was surprised to find the accompanying leaf of Darlingtonia. This was the first time that the plant
was found since it was first discovered by Mr Brackenridge in 1842 Dr Wilkes first collected it in flower]
J. Torrey

This flower a species of Sarracnia was
found by me in California near the Mud Volcano
in 49 & I have had it in an old memorandum
book ever since   C. Flint — it is of course old to you

129

The U.S. National Herbarium is part of the Smithsonian's National Museum of Natural History.

The New York Botanical Garden: An American Kew

Bookending the century that began with Lewis and Clark's western collections was the establishment of the New York Botanical Garden in 1891. The founders were Nathaniel and Elizabeth Britton, a husband-and-wife team of botanists. Nathaniel (1859–1934) was from a well-established Staten Island family. He studied geology at Columbia University and became a professor there, but his interests turned to botany. He assumed responsibility for the Torrey Herbarium, which had not been actively curated since Torrey's death in 1873. Elizabeth Knight Britton (1858–1934) grew up mostly in Cuba, where her family had a sugar plantation. She graduated from Normal (now Hunter) College in New York City and then taught there, during which time she developed a lifelong interest in bryology, eventually becoming the unofficial curator of the bryophyte collections at Columbia.

On a visit to the Royal Botanic Gardens at Kew while both were in their twenties, the Brittons conceived of the idea to build an American equivalent to Kew in New York City, an idea supported by fellow members of the Torrey Botanical Club.

▲ Elizabeth Britton, bryologist and co-founder of the New York Botanical Garden.

◀ Nathaniel Britton, specialist in the Cactaceae and the American flora, co-founder of the New York Botanical Garden.

Convincing the city to take on such a project was not easy, but Nathaniel Britton's family connections helped secure private funding for the purchase of the land in the Bronx, and the Brittons' dream came to life. The New York Botanical Garden herbarium opened its doors in 1901, with the Columbia University collection as its nucleus, supplemented by an impressive array of private herbaria, including that of Job Bicknell Ellis. The focus of the institution was to be on American plants and fungi, but not necessarily only on the plants of the United States. The Brittons believed that their country should be the leader in the exploration and documentation of plants of the entire western hemisphere. Acting upon that assertion, they carried out more than 20 years of exploration in the Caribbean region, sending 75 collecting expeditions there, yielding some 150,000 specimens.

The Brittons were not the only American botanists in their time to venture beyond national borders in pursuit of plant collections. From its inception in 1894, the botany program at the Field Museum focused on the tropics, developing one of the world's major collections of Central and South American plants. Cyrus G.

Interior of the New York Botanical Garden Herbarium, c.1908.

This building, opened in 2002, houses the William and Lynda Steere Herbarium and the LuEsther T. Mertz Library.

Pringle, one of the most prolific collectors and distributors of botanical specimens of his time, spent more than 26 years collecting plants in Mexico. Ynes Mexia, mentored by Alice Eastwood, collected widely in the American Southwest, Mexico, Chile, and the Amazon basin. Harvard's Arnold Arboretum, established in 1872, sent botanists Ernest H. Wilson and Joseph F. Rock on expeditions to China and Tibet to collect herbarium specimens and seeds.

To further decrease their dependence on European herbaria for comparative material, botanists with personal or institutional means often bought up European herbaria when they became available. The trend began with Tuckerman's purchase of Lambert's American specimens and Engelmann's purchase of the Bernhardi herbarium. Later in the 19th century, William Farlow and Roland Thaxter of Harvard University acquired the fungus collection of mycologist Narcisse Patouillard and the bryophyte herbaria of Thomas Taylor and Viktor Schiffner. Beginning in their Columbia University years and continuing after they established the New York Botanical Garden, the Brittons acquired the vascular plant herbaria of Carl Meissner and Otto Kuntze and the bryophyte herbaria of August Jaeger and William Mitten. All these herbaria were rich in type specimens, and most contained specimens from all over the world. Thus, by the early 20th century, American herbaria held sufficient quantities of recently collected and historical specimens to support most floristic work in the western hemisphere, except for types held in the major institutional European herbaria.

American Herbaria on a World Stage

Today the majority of herbaria and half of all herbarium specimens held in the United States are in institutions of higher learning, and indeed the history of herbarium growth closely follows the history of university education in this country. Between 1800 and 1850 more than 200 degree-granting institutions were established, focusing primarily on a classical and liberal arts education. Many of these institutions developed herbaria, usually thanks to the interests of individual professors; it was generally accepted that knowledge of local plants was part of a well-rounded education.

The Morrill Act of 1862 set in motion a plan by which the states could receive a portion of the profits derived from the sale of western lands by the federal government if they used the funds to establish programs of agricultural, mechanical, and military study as well as liberal arts. This led to the establishment of the so-called land-grant colleges. A second Morrill Act of 1890 continued the involvement of the federal government in funding programs in the land-grant colleges and also created

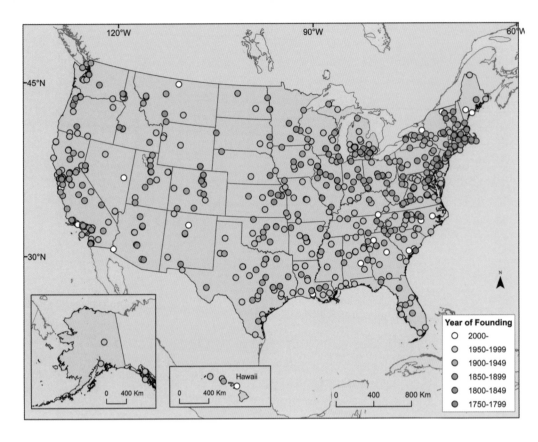

Map showing location of U.S. herbaria, points color-coded by year of founding.

the network of historically black colleges and universities (HBCUs). Most of the original land-grant colleges developed herbaria between 1860 and 1920. All still exist, although the University of Missouri's herbarium is now at the Missouri Botanical Garden, and the University of Maryland's herbarium is presently inactive for budgetary reasons. Among the HBCUs, six formed herbaria; the herbaria of Delaware State University, Howard University, and North Carolina A&T State University are the largest.

The Servicemen's Readjustment Act of 1944, more commonly known as the G.I. Bill, provided generous financial aid for veterans to attend college. This unprecedented educational opportunity and the post-war baby boom created a greater demand for seats in the college classroom than were available at the time. And global politics (especially Cold War concerns about keeping pace with the Soviet Union in technology) created the need for advanced programs in the physical and biological sciences as well as in political science and foreign languages. The result was a tremendous increase in the number of state-funded colleges, both four- and two-year, as well as an expansion of existing ones. Roughly 75% of American campus buildings were constructed between 1960 and 1985. This growth explains why the greatest increase in the number of herbaria in the United States took place between 1961 and 1980.

In the United States, unlike most of the other regions of the world, the proportion of herbaria and specimens held in federal institutions is low—only 10% of total specimens. Aside from the National Herbarium of the Smithsonian Institution, which holds the vast majority of federal specimens, herbaria are housed in the headquarters of land management agencies (e.g., national parks, national forests) as well as in some federal research institutions. Currently there are two tribal herbaria, the Navajo Nation Herbarium and the Nez Perce Tribe Herbarium. Two private corporations maintain herbaria: Amway Corporation, located in Ada, Michigan, and the McIlhenny Company (producers of Tabasco products) in Avery Island, Louisiana.

Herbarium Digitization

Beginning in the 1970s, U.S. herbaria began to enter data from specimen labels into electronic databases, transcribing basic collections metadata—such as the name of the organism and where, when, and by whom it was collected—into structured fields so that any of these terms could be used to search a collection. Since the physical specimens are stored by their scientific names, there had heretofore been no easy way to find specimens based on any other element of their associated data (e.g., all specimens collected by Elizabeth Britton, or all specimens from Ecuador). However, the early herbarium specimen databases were constrained by the absence of data standards for these elements. Processor speed and storage capacity were issues, too, along with the lack of user-friendly database interfaces. As the technology developed, however, most of these problems were solved. An international community of taxonomists, collections professionals, and computer specialists interested in biodiversity data formed the Taxonomic Databases Working Group (TDWG) in 1985 to support the development of standards for all types of natural history specimen data. One of these is the Darwin Core standard, which gives names and definitions for almost all the data elements one might find on a specimen label.

With the Internet and the Darwin Core, herbaria could share their databases with anyone in the world. High-quality digital photography made it possible to share images of the actual specimens as well as the transcribed label data. The nature of herbarium sheets (flat and mostly standard in size and shape) made them excellent candidates for rapid digitization.

Among the projects enabled by digitization was the Global Plants Initiative begun in 2003, funded by the Andrew W. Mellon Foundation under the direction of program officer William Robertson IV. The goal of this project was an online database of the type specimens from the world's herbaria. This groundbreaking project

advanced the study of plant diversity and the herbarium community in several ways. It provided data and images for more than 2 million type specimens housed in more than 300 herbaria worldwide, helping researchers find these critical specimens for biodiversity research. By providing funding for relatively simple but high-quality imaging technology as well as staff to do the work, many herbaria, especially those in developing countries, were empowered to digitize not only their type specimens but other specimens as well, and could share their holdings with a worldwide audience, thereby increasing the use of their herbaria. Funding for the Global Plants Initiative ended in 2015, but the project continues as new type specimens are added or discovered in collections. The database is now maintained by JSTOR (plants.jstor.org).

Digitization of herbarium specimens as well as other natural history collections advanced significantly in the United States between 2010 and 2020 through Advancing Digitization Biodiversity Collections (ADBC), a National Science Foundation funding program. This ten-year program has provided funding for consortia of institutions to digitize specimens that relate to particular research themes regarding environmental change and species interdependence. Through this program, many collections of algae, bryophytes, fungi, and vascular plants from most regions of the country have been digitized. The ADBC program has also provided funding for iDigBio, a central organizing hub for the digitization effort. A joint effort between the University of Florida and Florida State University, iDigBio hosts the national

data portal where all digitized data can be searched and provides training and workshops on various topics, further strengthening the herbarium community and creating stronger relationships with other types of natural history collections as well. As of February 2020, approximately 63 million specimen records for plants and fungi, most held in U.S. herbaria, are searchable through the iDigBio Portal (idigbio.org/portal).

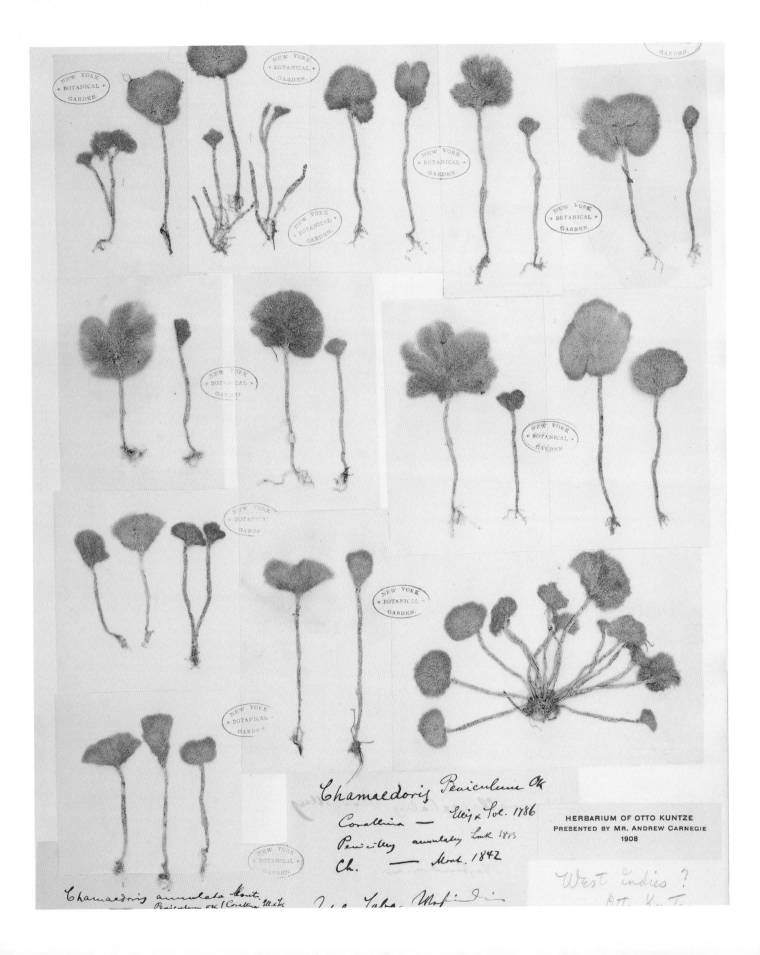

Chamaedoris Peniculum OK

Corallina ——— Ellis x Sol. 1786

Penicillus annulatus Lmk 1813

Ch. ——— Mont. 1842

HERBARIUM OF OTTO KUNTZE
PRESENTED BY MR. ANDREW CARNEGIE
1908

West indies ?

Ott. K. T.

Chamaedoris annulata Mont.

Peniculum OK (Corallina Ell. x Sol.)

Development
of Herbaria
Around *the* World

The European tradition of herbaria has spread to almost every country in the world, developing in a manner similar to that described for the United States: at first Europeans—explorers and emigrants to other countries—make plant collections for study by scientists back home, but as they become citizens and a national identity is formed, in-country herbaria become part of the scientific infrastructure for interpreting the history and natural resources of the new nation. The politics, traditions, and perspectives of each country give each herbarium origin story a distinct flavor, however. This chapter will explore how a national tradition of herbaria developed in selected countries around the world: Australia, Brazil, China, and South Africa.

◀ An herbarium specimen of
Chamaedoris peniculum, a green alga.

Australia

The sixth-largest country in the world and the only one to occupy an entire continent, Australia hosts over 30,000 species of plants, more than half of which grow nowhere else. Much of the continent is arid, though tropical rainforest can be found on the northeastern coast near Cairns, Queensland, and temperate forest and rainforest occurs in the southeast (mostly Victoria and Tasmania).

Early Exploration

James Cook's first voyage (1768–71) on the *Endeavour* set the stage for Australia's future as a British colony and later an independent nation. The expedition made landfall at 11 places on the eastern and northern coasts of Australia, and at each stop the ship's naturalists, Joseph Banks and Daniel Solander, made thousands of collections of plants, birds, fish, and reptiles. Cook named one of the expedition's eastern-coast landfalls Botany Bay for the large number of plants collected there. Overall more than 30,000 collections resulted from the expedition, including more than 1,300 new species of plants, and these became part of Banks' personal collection. Linnaeus suggested that the continent be called Banksia in honor of his role in documenting its distinctive biota, and although this did not happen, his name does grace one of Australia's most iconic genera, *Banksia*, of the protea family, named by Linnaeus' son. Banks' collections were illustrated as he collected them by Sydney Parkinson, a brilliant young artist who died of dysentery on the return voyage. Back in London, Banks hired other artists to create watercolor paintings from Parkinson's sketches to illustrate the flora of Australia he planned to write. Although the flora was never written, the exquisite illustrations of plants from all areas visited by the *Endeavour* were published in the series *Banks' Florilegium* (1980–90).

A French expedition led by Bruni D'Entrecasteaux aboard the vessel *Recherche* visited Tasmania and southeastern Australia in 1792 and 1793, with botanist Jacques Labillardière (1755–1834) aboard as the voyage naturalist. Labillardière made many collections in Australia, which he documented in his *Novae Hollandiae Plantarum Specimen* (1804–06). This two-volume work

Banksia serrata (old man banksia, saw-tooth banksia), a species collected by Banks and Solander in Botany Bay. This tree occurs along the eastern coast of Australia.

▲ A possible type specimen of *Lagenophora stipitata*, in the sunflower family, collected (as *Bellis stipitata*) by Jacques Labillardière in Australia.

▶ Type specimen of *Banksia neoanglica* (originally described as a variety of *B. spinulosa*), showing not only the foliage and dense flowerhead but also the distinctive woody fruits that characterize the genus.

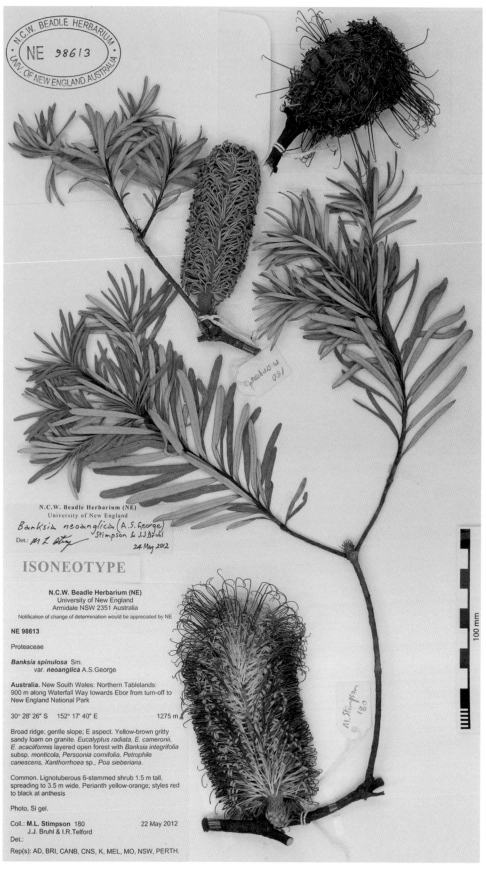

N.C.W. Beadle Herbarium (NE)
University of New England

Banksia neoanglica (A.S.George)
Det.: *M. L. Stimp* Stimpson & J.J.Bruhl
24 May 2012

ISONEOTYPE

N.C.W. Beadle Herbarium (NE)
University of New England
Armidale NSW 2351 Australia
Notification of change of determination would be appreciated by NE

NE 98613

Proteaceae

Banksia spinulosa Sm.
var. ***neoanglica*** A.S.George

Australia. New South Wales: Northern Tablelands:
900 m along Waterfall Way towards Ebor from turn-off to
New England National Park

30° 28' 26" S   152° 17' 40" E   1275 m

Broad ridge; gentle slope; E aspect. Yellow-brown gritty
sandy loam on granite. *Eucalyptus radiata, E. cameronii,
E. acaciformis* layered open forest with *Banksia integrifolia*
subsp. *monticola, Persoonia cornifolia, Petrophile
canescens, Xanthorrhoea* sp., *Poa sieberiana.*

Common. Lignotuberous 6-stemmed shrub 1.5 m tall,
spreading to 3.5 m wide. Perianth yellow-orange; styles red
to black at anthesis

Photo, Si gel.

Coll.: **M.L. Stimpson** 180   22 May 2012
J.J. Bruhl & I.R.Telford
Det.:

Rep(s): AD, BRI, CANB, CNS, K, MEL, MO, NSW, PERTH.

would never have been published without the help of Joseph Banks. War broke out between France and Britain while the expedition was on its voyage home, and at a resupply point in Java, the British captured the *Recherche*. The ship and crew were imprisoned and the collections were confiscated, destined for London as war booty. Demonstrating that the bond between botanists sometimes transcends nationality, Banks interceded to have the collections returned to his French colleague. Most of Labillardière's specimens are housed in the herbaria of the natural history museums in Paris and Florence. Labillardière is eponymized in *Billardiera*, a genus of woody vines known only from Australia.

Banks also advised the 1801–03 expedition led by Matthew Flinders, whose objective was a coastal survey of Australia. Robert Brown, the ship's botanist, found ample opportunities for plant collection on this trip, amassing more than 6,000 specimens, many of which represented new species. Bungaree, an Australian native of the Kuringgai people from the Broken Bay region north of Sydney, traveled with the expedition during the circumnavigation of the continent. He assisted with navigation and communicated with the native people encountered on the voyage. He also helped Brown to record Aboriginal names for plants in his diary, providing an important resource for Australian botany and Aboriginal linguistics. Bungaree later assisted several other Australian explorers.

Upon return to England, Brown began work on an Australian flora; however, the first volume of the series, *Prodromus Florae Novae Hollandiae et Insulae Van Diemen* (1810), sold poorly, and he did not continue it. He became librarian to Joseph Banks and inherited Banks' herbarium and library after his death in 1820. When the collection was transferred to the Natural History Museum, Brown was appointed its keeper. He continued research on plants for the remainder of his career and is now best remembered for Brownian motion, the random motion of particles in a fluid, which he first observed in pollen grains. His Australian specimens are deposited at the Natural History Museum in London, with duplicates in other major herbaria around the world.

Amalie Dietrich was a later and very different European explorer of Australia. Born in 1821 in Germany, she had only a very basic formal education, but when she married Wilhelm Dietrich, a natural history collector, she learned his trade and in fact far surpassed him in the breadth and quality of her collections. As a team, the Dietrichs collected in the Alps for specimens that they sold to chemists (for use in medicines) and museums. Wilhelm lost interest in the work, however, leaving the responsibility for collecting to Amalie; in return, he was supposed to maintain their home and care for their daughter, Charitas. He failed to hold up his end of the bargain and left Charitas on her own for extended periods while he had affairs and

▶ Portrait of Amalie Dietrich by Christian Wilhelm Allers, 1881.

▶▶ Type specimen of *Pagetia dietrichiae* (now *Bosistoa medicinalis*), bearing a photocopy of Dietrich's original specimen label and obtained by the New York Botanical Garden from the herbarium of Hamburg University, which incorporated Dietrich's botanical collections from the Godeffroy Museum.

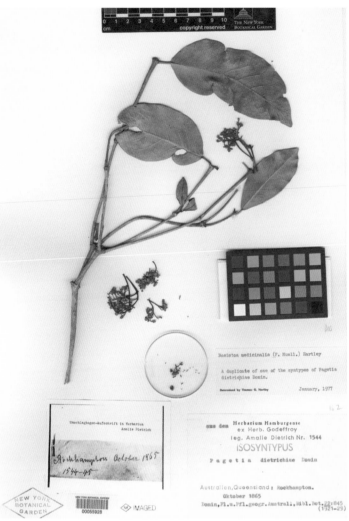

otherwise squandered the meager family funds. Eventually Amalie left him, taking Charitas with her, to start a solo life and career as a collector. Johann Godeffroy, a German shipping magnate who was building a private natural history museum in Hamburg, hired her to collect in Australia. The terms of her contract were difficult for the mother of a dependent child—she was to spend ten continuous years collecting specimens in Australia, and her daughter could not accompany her.

Amalie arranged for her daughter to live with the family of a fellow collector and left for Australia in 1863. She collected alone, primarily in Queensland, amassing not only specimens of plants, insects, and vertebrates but also the skeletal remains of Aboriginals, all of which were sent to Godeffroy for his family's museum. Her specimens of Australian wood won a prize at the Paris International Exhibition of 1867, and she continued to sell some of her more than 20,000 plant collections, including algae and bryophytes, to Kew and other herbaria. Upon her mother's early return from Australia in 1872, Charitas was shocked by her appearance: she was shabbily attired and the years of exploration had aged her. Dietrich was not broken by her experience, however. Widely recognized for the excellent quality of her natural history specimens, she was hired to work at Godeffroy's museum in Hamburg, living above it until the contents were donated to the Hamburg Museum, at which time she became curator of that museum. Amalie Dietrich passed away in 1891. She did not publish on her collections, but more than 30 species were named for her, based mostly on her specimens.

Founding Botanists

It was Joseph Banks who suggested that Australia would be a suitable place for the relocation of convicts in order to ease overcrowding in British prisons. Accordingly, the government established a penal colony there in 1788, in Botany Bay, where the city of Sydney would later develop. Voluntary settlers came as well, many intending to support themselves using the same farming practices they used in Europe. Such methods did not always succeed in the drier, hotter Australian climate, however, and when crops failed, settlers turned to the native flora and fauna for food, which came to be known as bush tucker, a corruption of *tuckout*, a general Aboriginal pidgin word for food. Aboriginal people shared their knowledge of wild edible plants, both greens and starchy roots, with the settlers. One of the plants that sustained settlers in difficult times was *Tetragonia tetragonoides* (New Zealand spinach), which is native in the Sydney region. Used by members of Cook's expedition to ward off scurvy and grown extensively as a green vegetable by settlers, this plant was eventually introduced into Europe and now is a worldwide agricultural crop.

Most early Australian settlers therefore developed some interest in the native Australian flora as part of their survival strategy. Some, however, like James Drummond, took a strong interest in documenting the flora of their new land—and supplemented their incomes by shipping these unfamiliar plants to growers and botanists in Europe. Born in Scotland in 1786, Drummond was curator at the Cork Botanical Gardens in Ireland from 1809 until it closed, due to lack of funds, in 1828; the following year he and his family emigrated to Australia, where they were early settlers along the Swan River, near present-day Perth. The Drummonds later moved farther inland to the Avon Valley region.

Appointed to an unpaid position as government naturalist shortly after arriving in Australia, Drummond gained renown for the quality and quantity of his collections, despite the fact that paper for pressing plants was a rare commodity—sometimes

CLOCKWISE, FROM TOP LEFT
A grass tree (*Xanthorrhoea*), growing near Glen Innes, New South Wales, Australia.

Type specimen of *Grimmi cygnicolla*, a moss collected by James Drummond in 1843.

Drummond is eponymized in the Australian endemic *Marsilea drummondii* (common nardoo). Aboriginal people ground the dried spore pods of this aquatic fern into a powder that was then mixed with water to form an edible dough.

he apparently had to resort to drying specimens between layers of grass tree leaves (members of the genus *Xanthorrhoea*). Drummond collected widely in Western Australia, supported largely through his relationship with William Hooker of Royal Botanic Gardens, Kew, who arranged for the purchase of Drummond's seed and plant collections. Drummond described several plant species in publications by Hooker and others, and over 100 species were named for him. His main herbarium is deposited at the Royal Botanic Gardens in Melbourne.

The figure that looms largest in 19th-century Australian botany is Ferdinand von Mueller, born in 1825 in Rostock, Germany. He was in training to become a physician when most of his family died from tuberculosis. Following this loss, he and his surviving sisters decided to emigrate to Australia, arriving in Adelaide in 1847 but later settling in Melbourne. He began to collect plants, partly out of interest in documenting a still largely unknown flora but also for the possibility of discovering new medicines, perhaps motivated by his own family tragedy. On recommendation from

William Hooker, he became the first government botanist in Victoria, a post he held for 43 years concurrent with some other appointments, including the directorship of the Royal Botanic Gardens in Melbourne from 1857 to 1873.

Von Mueller was a prodigious collector, participating in a wide range of expeditions across the continent. Realizing the documentation of the Australian flora was far too large a job for one person, he established a network of plant collectors throughout Australia—in exchange for specimens, he sent them packets of garden seeds. He advertised for collectors in several newspapers across Australia. More than 200 women collectors were part of his network, and von Mueller frequently named new species for them. Von Mueller's plant collections formed the basis for the National Herbarium of Victoria, still the largest herbarium in Australia.

Von Mueller wrote as vigorously as he collected, producing 11 volumes of observations and descriptions of new species in the series *Fragmenta Phytographica Australiae*, two volumes on the plants of Victoria, and treatments of *Eucalyptus*, *Acacia*, and other large and complicated Australian genera. He wanted to write a flora of the continent but was dissuaded by Hooker and his fellow Kew botanist George Bentham because they didn't believe he could do a comprehensive treatment without reference to the historical collections in European herbaria. Instead they proposed that von Mueller send his entire herbarium, over a period of years, to Kew for examination. Because he served at the pleasure of the British government, he had no choice but to comply. Von Mueller was acknowledged in the resulting publication, *Flora Australiensis*, but the authorship was solely that of George Bentham, a bitter disappointment to von Mueller. The German government recognized his contributions to the understanding of the Australian flora by conferring the title of baron upon him, and he is commemorated in the names of many plants and Australian places, such as the Mueller Ranges in Western Australia and the Mueller River in Victoria. *Muelleria*, the journal of the Royal Botanic Gardens Victoria, begun in 1955, is named in his honor.

Despite the exclusion of von Mueller, Bentham's *Flora Australiensis* was a masterful work. It included descriptions for 8,125 species, and its publication, between 1863 and 1878, marked the end of the preliminary phase of Australian plant exploration. Afterward, study of the flora became more localized, leading to the development of multiple centers of botanical expertise across the country. The next 50 years would

▲ Ferdinand von Mueller (1825–1896), Australia's most prolific 19th-century plant collector.

▶ Type specimen of *Hakea brookeana* (now *H. obliqua*), named by von Mueller for botanical collector and illustrator Sarah Brooks, who sent him specimens from the Israelite Bay region of the Western Australia.

MEL 108085

MEL108085

LECTO-
TYPE

NATIONAL HERBARIUM OF
VICTORIA (MEL), AUSTRALIA

Photographed June 21, 1985

HOLO-TYPE

PHYTOLOGIC MUSEUM OF MELBOURNE.

Hakea Brookeana
F.M.
Israelite Bay
fruit reminds of H. platysperma
1885.   Miss S. Brooke
BARON FERD. VON MUELLER, PH. & M.D.

"At or towards Mt. Ragged" - see TYPE description;
Aust. Journ. Pharm. I, p. 450 (1886)

see the growth of herbaria in all the state capitals, often in association with botanic gardens. For many years, herbarium collections made in the Northern Territory were maintained in Canberra, but eventually herbaria were established there in Alice Springs (1954) and Darwin (1967).

The first state flora was for South Australia, published by John McConnell Black between 1922 and 1929. Eventually each state except Western Australia had a flora of its vascular plants, but research on cryptogamic groups proceeded more slowly. The first attempt at a fungus flora for any large area of the country was made by John Burton Cleland, Black's contemporary and colleague, who published *Toadstools and Mushrooms and Other Larger Fungi of South Australia* in 1934–35. Bryan Womersley of the University of Adelaide described some 320 new genera and species of algae and produced a major compendium of the marine algae of southern Australia in six volumes between 1984 and 2003. There have been two regional works on mosses: *The Mosses of Southern Australia* (1976) by Ilma Stone of the University of Melbourne and George A. M. Scott of Monash University, and *Mosses of South Australia* (1980) by David Catcheside. The collections of Black, Cleland, Womersley, and Catcheside are deposited at the State Herbarium in Adelaide, and those of Stone and Scott are maintained at the University of Melbourne.

ARAUCARIA MÜLLERI AD. BRONG. & A. GRIS

Chrom. P. De Pannemaeker.   *J. Linden, publ.*

Araucaria muelleri, native to New Caledonia. From C. A. Antoine, *L'Illustration Horticole* (1882), vol. 29, plate 449.

Modern Documentation and Herbaria

Emphasis on regional floras was a natural choice for Australian botanists, given inherent difficulties such as a lack of roads in the interior, long periods of drought, and the relatively few scientists available to do this work. Eventually, though, it became clear that a complete understanding of plant groups required a continental perspective. The Australian Biological Resources Study (ABRS) was established in 1973 to promote the study of Australian plants and animals with the goal of producing nationwide treatments of Australian flora and fauna. The *Flora of Australia* project

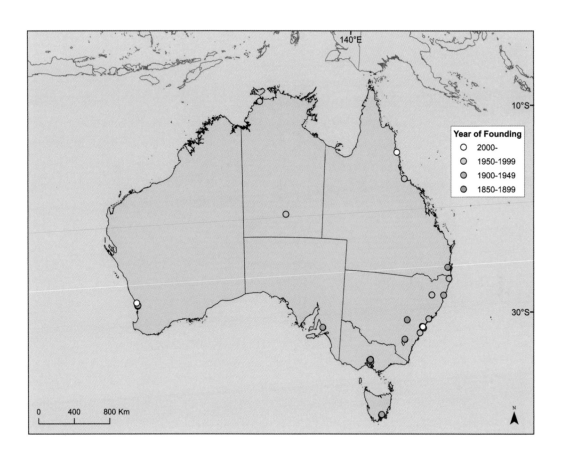

Map showing location of Australian herbaria, points color-coded by year of founding.

was launched with considerable fanfare at the 13th International Botanical Congress, held in Sydney in August 1981. Sixty hardcopy volumes of the series were published between 1981 and 2015; the flora continues now as a digital resource.

Today Australia's 36 herbaria hold approximately 8.4 million specimens. Most specimens (80%) are held in government institutions, primarily botanic gardens. Sixteen herbaria are associated with universities. An important organization enabling the study of the plants and fungi of Australia and New Zealand is the Council of Heads of Australasian Herbaria (CHAH), formed with the goal of creating a forum for discussion of herbarium- and flora-related issues. At first the group included only representatives of government herbaria but eventually welcomed all herbarium representatives, as the Council of Australasian Herbaria.

CHAH was instrumental in the creation of the Australian Plant Census (APC), a list of accepted names for the country's plants; the list is maintained by consensus of the Australian botanical community, mediated by CHAH. Australia's Virtual Herbarium (AVH), begun in 2001 with commonwealth, state, and private funding, is another cooperative project among Australia's herbaria fostered by CHAH; Australian herbaria agreed to digitize the holdings in their collections for the project and continue to contribute these to a single searchable database. Both the AVH and the *Flora of Australia* contribute to the Atlas of Living Australia (ALA), a portal to digitized information about all Australia's biota. With a broad funding base and strong relationships to other national and international organizations and biodiversity data sources, the ALA has set the standard for how nationwide resources can be shared, with the goal of improving the understanding of biodiversity while also helping to develop policies to sustain it. Attentive to the needs of a non-technical audience, the ALA provides a user interface and content to facilitate use of these data by educators and students as well as by the general public.

Despite an admirable level of attention to and support for herbaria, Australia was the scene of a tragic destruction of herbarium specimens in 2017. The Queensland Herbarium requested specimens of the genus *Lagenophora*, a member of the sunflower family, from the National Herbarium in Paris for study by one of their scientists. In response, 105 specimens were sent on loan. The shipment included six type specimens, including at least one collected by Labillardière on the D'Entrecasteaux voyage. Instead of being delivered to the Queensland Herbarium, the specimens were incinerated by customs authorities. Because of the devastating effect of invasive species on the native biota of Australia and on Australian agriculture, the nation maintains very strict quarantine measures, which require extensive paperwork and precise labeling of parcels, and stipulate the destruction of any that are insufficiently identified as safe for entry into the country. A series of miscommunications led these invaluable herbarium specimens to fall into the category of threat to the nation.

The destruction of these herbarium specimens received wide attention in the Australian and international press, and senior Australian diplomats made a visit to the Natural History Museum in Paris to apologize in person to the museum's president. Horrified by the incident, CHAH has worked extensively with the Australian Department of Agriculture to put safeguards in place to prevent such occurrences in the future.

Brazil

Brazil hosts approximately 46,000 plant species, more than twice as many as the slightly larger United States. The Amazon rainforest still occupies approximately 40% of the country's total area; other biodiversity-rich habitats include the savanna-like cerrado of the southern central region and the Atlantic Forest, rich

Legno del brasile (brazil-wood) by Felice Cassone (1815–1854), from his *Flora Medico-Farmaceutica* (1847–52).

in endemics and now highly fragmented along the populous coastline. The country gets its name from pau-brasil (aka brazilwood, pernambuco), a tree of the genus *Paubrasilia* endemic to the Atlantic Forest.

Early Exploration

The Portuguese established their first colony in Brazil in 1532, settling mostly in coastal areas and swiftly developing a sugarcane industry, for which African slaves provided most of the workforce. Although the Portuguese claimed all coastal Brazil as a colony, in 1630 the Dutch took control of northeastern Brazil and established their capital along the coast in Recife, in the state of Pernambuco. Johan Maurits, the settlement's administrator, established the first European-style garden in Brazil and also provided support for the first known biological exploration, by German naturalist and astronomer Georg Marcgrave (1610–1644) and physician Willem Piso. They arrived in 1638 and spent six years exploring Brazil, later traveling to the islands of the Caribbean and finally to Africa, where Marcgrave became ill and died. *Historia Naturalis Brasiliae* (1648) contains an illustrated account of the natural history observations of both explorers. Conversion of Marcgrave's notes into text for the volume was challenging—he recorded his notes in code, purportedly to keep Piso from appropriating his work. Marcgrave's pioneering work is remembered in *Marcgravia* and the Marcgraviaceae, a family of neotropical woody shrubs, vines, and epiphytes.

In the last quarter of the 18th century, the Portuguese crown began to promote natural history study in Brazil, partly driven by the Enlightenment's quest for knowledge but also by a desire for more products for trade. The sugarcane industry, once booming, now faced competition from the Caribbean, and the placer gold that had been discovered in southeastern Brazil in the 1690s was exhausted. To maintain Portugal's strategic advantage in the region, Italian naturalist Domenico Vandelli, a correspondent of Linnaeus and professor at the University of Coimbra, was charged with organizing natural history expeditions to Brazil, from which specimens, carefully guarded as state secrets, were to be brought back for study. One of those he engaged in this work was Alexandre Rodrigues Ferreira (1756–1815), who had obtained baccalaureate and doctoral degrees from the University of Coimbra.

▲ The dense orange-red heartwood of pau-brasil, commonly used to make inexpensive bows for stringed instruments, also yields a red dye for fabrics and wool.

▶ Marcgrave's *Herbarium Vivum Brasiliense*, a volume of 177 plants collected during his time in Brazil, is preserved in the Natural History Museum in Copenhagen. This set likely includes the first dried specimens of plants from tropical America.

HERBARIUM VIVUM
BRASILIENSE.
Plantarum et Fructuum.
â
Viro Clariffimo
Dño. Georgio Marcgravio
de Liebftad. Mifnic. German:
In Brafilicana Infula fingu
lari ftúdio Colle.
ctorúm et affer.
vatorúm.

186. 150.

Ferreira was born in Bahia, Brazil, the son of a Portuguese merchant. In 1783 he left on a remarkably ambitious expedition that began in Belém, at the mouth of the Amazon; he traveled upriver to the border with Colombia, then later followed the Madeira and Guaporé Rivers to Mato Grosso before returning to Belém nine years later. Arriving home to Lisbon in 1793, Ferreira dutifully turned collections over to Portuguese scholars, who did nothing with them. Half his specimens were eventually taken to Paris as war booty by Napoleon's forces, where they were turned over to the Natural History Museum. Ferreira became a bureaucrat, working in various administrative capacities in Portuguese government, and his notes and watercolor paintings from his expedition were never published. Today Ferreira's known specimens are approximately split between the herbaria of the natural history museums in Lisbon and Paris. Portugal's secretive approach to natural history exploration yielded them few returns; other European botanists did describe a number of Ferreira's plants as new species, but these were based mostly on those maintained at the National Herbarium in Paris.

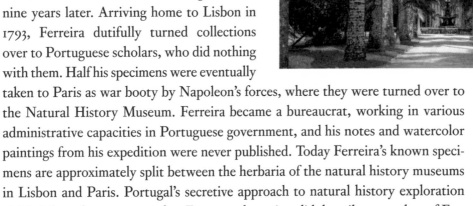

An allée of palms at the Rio de Janeiro Botanical Garden, Brazil's first, founded in 1808 by João VI of Portugal.

Scientific exploration in Brazil became more open in the early 19th century, when the Portuguese monarch João VI came with his court to live in Rio de Janeiro, fleeing Napoleon's invasion of Lisbon. João welcomed scientific exploration by all. He established Brazil's first botanic garden in 1808, and the first natural history museum in 1818, both in Rio. He also encouraged settlers, who by establishing roads and farms opened up the interior, leading to the domination or elimination of indigenous people in some areas. João raised Brazil to the status of a "united kingdom" with Portugal before returning there in 1821. His son Pedro stayed behind to govern the semi-autonomous Brazil, but a year later declared Brazil's independence and gave himself the title of emperor, becoming Pedro I of Brazil.

This political climate was the backdrop for the expedition of Augustin Saint-Hilaire (1779–1853). Raised in Orléans, France, Saint-Hilaire spent the years 1816–22 and 1830 and traveled some 5,500 miles exploring central and southern Brazil. He collected 24,000 specimens of 6,000 plant species, and also birds, insects, and mammals. New York Botanical Garden botanist William Wayt Thomas, who has seen Saint-Hilaire's original field books, describes them as "amazingly detailed, with complete descriptions of the specimens and species he collected." Saint-Hilaire also provided original data on the geography and culture of Brazil, which makes his

Type specimen of *Asteranthos brasiliensis*, in the Brazil nut family, collected by Alexandre Rodrigues Ferreira and named by Parisian botanist R. L. Desfontaines.

works primary sources for historians as well as botanists. Two major works resulted from this journey: *Flora Brasiliae Meridionalis* (1825–33), Brazil's first flora, and a travelogue, *Voyages dans l'Intérieur du Brésil*. His collections were deposited at the National Herbarium in Paris, not maintained as a separate unit but dispersed among the general herbarium collections; some 8,900 of Saint-Hilaire's plant specimens are held there, and duplicates are deposited in other herbaria.

A contemporary of Saint-Hilaire's and perhaps the greatest 19th-century European explorer in Brazil was Philipp von Martius (1794–1868). Born in Erlangen, Germany, he traveled to Brazil with zoologist Johann Baptist von Spix as part of the wedding party of the Maria Leopoldina of Austria, who would marry Pedro I—the naturalists were part of her dowry arrangement! Departing from Rio in December 1817, Martius and Spix traveled overland north to Belém, arriving in July of 1819. They spent the next nine months exploring the Amazon, with Martius venturing as far as the Rio Caquetá in Colombia.

The approximately 15,000 botanical specimens Martius collected on this trip became the basis for many future botanical studies, especially the monumental *Flora*

▲ Isotype of *Viola cerasifolia*, a violet, collected by Saint-Hilaire and obtained from Paris by the New York Botanical Garden.

▼ Augustin Saint-Hilaire by Brazilian artist Henrique Manzo (1896–1982).

▶ Specimen of *Swartzia apetala* var. *blanchetii*, in the legume family, collected in Brazil by Martius.

c
d

REVISION OF SWARTZIA (LEGUMINOSAE)

Swartzia apetala var. blanchetii (Bentham) Cowan

Richard S. Cowan   1963

Swartzia apetala
(Raddi) Imkut

Martii Herber Florae
Brasil No 654

Brasiliensis. Martius was the original editor of this work, the first volume of which was published in 1840; it would take two subsequent editors and the contributions of 75 other botanists to see the work through to the final volume, the 40th, published in 1906. The work treated 23,000 species, or approximately half of today's known flora of Brazil. Separately, Martius described 70 genera and 400 species in his *Nova Genera et Species Plantarum Brasiliensis* (1824–32) and hundreds of palms in his *Historia Naturalis Palmarum* (1823–53). Although he served for years as the keeper of the botanic garden and herbarium in Munich, where the specimens from his Brazil trip were deposited, Martius also maintained a private herbarium. This *Herbarium Martii*, as it is known, contained some 300,000 specimens, making it one of the largest private herbaria assembled. The Belgian government acquired this collection after Martius' death, and it became the nucleus of the herbarium at the Meise Botanic Garden.

Engraving of the forest on Mt. Corcovado, near Rio de Janeiro, after the original watercolor by Austrian artist Thomas Ender (1793–1875). From Martius' *Flora Brasiliensis* (1840), vol. 1, part 1.

Founding Botanists

As knowledge began to accrue about Brazil's natural and human history, the young country needed an organizational structure to conserve and share knowledge and direct future work. In 1838, naturalists, botanists, and other scholars gathered in Rio to create the Instituto Histórico e Geográfico Brasileiro (IHGB) to address this topic. The group held regular meetings and published a journal. They also organized competitions, promoted scientific expeditions, and gathered colonial-era documents, spread throughout Europe and elsewhere in Brazil, that were relevant to their country's history. Their emblem was a branch of the pau-brasil, recognizing the importance of Brazil's natural resources to the country's history.

In 1840, IHGB sought advice from scholars in Europe and Brazil about how to approach the writing of Brazil's history, offering a prize to the essay with the best plan. It was Martius that won the prize! His essay proposed that Brazil organize historical studies around the country's three major ethnic groups (European, African, indigenous), a process that would include study of language and culture as well as geography, natural history, paleontology and geology. Following this advice, the country began a period of exploration by Brazilians, yielding many natural history collections and human artifacts, many of which were deposited in Rio de Janeiro's National Museum.

João Barbosa Rodrigues (1842–1909) was one of Brazil's greatest botanists, known especially for his work on orchids and palms. His father was a Portuguese merchant, and his mother was indigenous Brazilian. An accomplished poet and artist, he received a classical education and was fluent in Portuguese and French. He began his botanical expeditions in 1868 with trips within the states of Minas Gerais (where he was raised) and Rio de Janeiro, extending later to Ceará, Paraíba, Pernambuco, Pará, and Espírito Santo. He was commissioned by the Brazilian government in 1871 to explore the Amazon basin, focusing on palms. Rodrigues lived there with his family for several years, collecting Indian artifacts and animals as well as plants.

When this expedition ended, Rodrigues could find no continuing support for his exploration. His godmother, who happened to be Princess Imperial Isabel, granddaughter of Pedro I and heir to the Brazilian throne, came to his rescue. In 1883 she provided the funding for him to create the Museu Botânico do Amazonas in Manaus. The short-lived museum, with Rodrigues as director, had botanical, ethnographic, and chemical sections, but in 1890 Rodrigues was offered the directorship of the Rio de Janeiro Botanical Garden, and after he left Manaus, the museum was not maintained.

Rodrigues found the Rio de Janeiro Botanical Garden in a poor state, with no library or herbarium, and many of the living collections unlabeled. Here his royal connections came in handy again. Pedro II (Isabel's father) helped Rodrigues improve the garden's resources, although by this time Pedro had been deposed as emperor and was living in Europe. To form the nucleus of the garden's herbarium, Pedro donated the herbarium of Antoine Fée (1789–1874), professor of botany at the University of Strasbourg, who was a specialist in Brazilian ferns and had amassed a collection of 25,000 specimens in the course of his career. Building on this base, Rodrigues built the herbarium at the Rio de Janeiro Botanical Garden into the largest in the country.

While in Manaus, Rodrigues had begun work on an ambitious project of color illustrations of living orchids, prepared with the help of his third wife, Constança. The set ultimately contained 900 plates, many of which illustrated species that Rodrigues had discovered. Unable to find financing to publish the color plates, he described some 540 new orchids without illustrations in a two-volume work, *Genera et Species Orchidearum Novarum* (1877–81). He allowed another botanist, Alfred

Cogniaux, to publish black and white versions of some of his colorful paintings in Cogniaux's treatment of orchids for *Flora Brasiliensis*. Disappointment at not being able to publish his orchid paintings stayed with Rodrigues the rest of his life; they were finally published in *Iconographie des Orchidées du Brésil* in 1996.

Rodrigues had better luck publishing his work on palms, but here too the challenges of doing serious systematic work on plants outside of Europe in his time are evident. His *Sertum Palmarum Brasiliensium* (1903) treated 382 species of palms in 42 genera, 166 of which were described as new. This time the Brazilian government helped to finance the publication, and it included 174 of his magnificent watercolors. The impact of this work was lessened somewhat through the actions of Scottish palm specialist James W. H. Trail, who had accompanied Rodrigues on an expedition to the Trombetas River in the state of Pará. Rodrigues had allowed Trail to study

▲ Rodrigues' collections on display in the Rio de Janeiro Botanical Garden's museum.

▶ *Cattleya elongata* from Rodrigues' *Iconographie des Orchidées du Brésil.*

Vol. 4: t. 48   Cattleya elongata Barb. Rodr.

311

his drawings and notes, and sometime after the publication of *Sertum Palmarum*, Trail published some of the same species with different names, relegating Rodrigues' names to synonymy. Trail's actions disregarded the principle of botanical nomenclature that gives the first published name priority, and so Rodrigues was especially dismayed when later German botanist Carl G. O. Drude, who had authored the treatment of palms for *Flora Brasiliensis* (1881–82), used Trail's names instead of his.

Despite this disregard of Rodrigues' work by European contemporaries, his great contributions to Brazilian botany are well recognized. The Rio de Janeiro Botanical Garden's museum was named for Rodrigues on the 100th anniversary of his birth, and he has been eponymized in three plant genera—*Barbosa* (now *Syagrus*, a palm), *Barbosella* (an orchid), and *Brodriguesia* (a legume)—as well as in the botanical journal *Rodriguésia*.

Modern Documentation and Herbaria

For nearly 70 years after Rodrigues' death in 1909, documentation of Brazil's plant life remained a major and increasingly urgent concern of the Brazilian government, as the country's flora was lost to land development. In 1976 the Brazilian National Research Council initiated Programa Flora, with the aim of making a detailed inventory of Brazil's vegetation, through collecting programs and creation of a computerized database of information derived from herbarium specimens. These efforts would be a prelude to the establishment of regional research centers throughout Brazil to carry out local floristic inventories, documentation of economically useful plants, identification of ecological problems, and development of conservation strategies. The program would also stimulate the education and training of Brazilian botanists, especially in plant systematics and data management, through graduate courses and short-term training programs.

The ambitious program was divided into five regional projects, the first of which to launch was Projeto Flora Amazônica. A shortage of botanists and computer experts in Brazil prompted that country to seek American help for the project. The U.S. National Academy of Sciences sponsored an initial meeting in Brasília attended by both American and Brazilian representatives, which paved the way for collaboration in the project, and the National Science Foundation provided support for the participation of 55 non-Brazilian participants in 25 expeditions over eight years, in which they collaborated with 36 Brazilian botanists.

Ghillean T. Prance (b.1937) led the American participation in Projeto Flora Amazônica. Born in England, Prance received his Ph.D. from Oxford, and afterward came to the New York Botanical Garden, beginning as a postdoctoral fellow. He was later appointed to a regular position as B. A. Krukoff Curator of Amazonian Botany, which afforded him the opportunity to spend long periods of time in the Amazon. At one point he spent two years in Manaus establishing a graduate training program at the National Institute of Amazonian Research, in addition to conducting field research; he was Director of Research at the New York Botanical Garden when Projeto Flora Amazônica began.

Computers were to be a key component of Projeto Flora Amazônica. Specimen label data would be submitted to a central processing center where it would be transferred into machine-readable form using the TAXIR (TAXonomic Information Retrieval) system, a software product developed in the United States but configured for installation in Brasília. TAXIR, written in FORTRAN IV and deployed on a mainframe computer, was one of the only software systems available for a project of this magnitude; however, the vision for the computerization component of the project was a bit ahead of the technical capabilities of the time and was never fully realized. In other aspects, however, the project was a great success, resulting in collections of approximately 33,000 vascular plants and 16,000 bryophytes and fungi. Projeto Flora Amazônica provided field experience in the Amazon for a cohort of Brazilian and foreign botanists, facilitating future collaboration among them and the next generation of field and laboratory research.

Projeto Flora Amazônica also paved the way for later large-scale initiatives to document the Brazilian flora. The targets of the Convention on Biodiversity's Global Strategy for Plant Conservation (GSPC) were renewed and updated in 2010, influencing government-funded research efforts in countries that are signatories to the convention. Brazil's response was the *Catalog of Plants and Fungi of Brazil*, published and launched online that same year. Achieving this milestone required the commitment of over 23 person-years by more than 400 Brazilians and foreign scientists. Building on the *Catalog* is the Brazilian Flora 2020 project, aiming to contribute to the GSPC's goal of an online flora of the world by that year. This project is engaging nearly 700 scientists, from Brazil and elsewhere, in the assembly of descriptions, identification keys, and illustrations for all species of plants, algae, and fungi known in the country. Another component of the project is the virtual repatriation of specimens collected in Brazil that are deposited in herbaria in other countries. Brazilian students, usually enrolled in graduate programs in botany at Brazilian universities, have received government funding to spend periods of time at the herbaria of the New York Botanical Garden, the National Herbarium in Paris, and others. The students capture images and data from Brazilian specimens, which are then transferred to Brazilian databases.

Projeto Flora Amazônica
The New York Botanical Garden
Museu Amapaense Angelo Moreira da Costa Lima
Museu Paraense Emílio Goeldi

No. 17695   Pteridophyte

        Trichomanes diversifrons (Bory) Sadeb.

Brazil. Amapá: Município de Macapá, vicinity of
Serra do Navio, 7 km NNW of village of Serra do
Navio on road to Água Branca. 1°3'N, 52°4'W.
Non-inundated moist forest.

Growing on rock along stream.

S. Mori, R. Cardoso &   4 Jan 1985
R. Souza
Fieldwork supported by NSF (U.S.A.) & CNPq (Brazil)

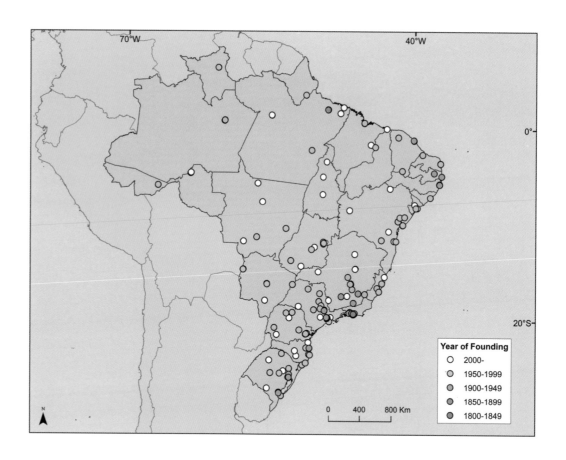

Map showing location of Brazilian herbaria, points color-coded by year of founding.

◄ *Trichomanes diversifrons* (filmy fern), a specimen from one of the many Projeto Flora Amazônica expeditions, collected in the Brazilian state of Amapá in 1985.

Today Brazil has 180 active herbaria, with a total of 8.4 million specimens. The oldest herbaria (founded in 1831 and 1890, respectively) are found in Rio, the first at the National Museum and the second at the Rio de Janeiro Botanical Garden. Recent growth has been very strong—the number of herbaria and specimens has almost doubled since 2000. Herbaria are found in all states of Brazil, with the strongest concentration in the southern states of Rio de Janeiro and São Paulo. The majority (68%) of Brazilian herbaria are associated with universities, followed by research institutes (16%) and botanic gardens (10%); the rest are housed at museums, parks, and nature reserves. Specimen data from many of Brazil's herbaria are shared through Brazil's Virtual Herbarium of Flora and Fungi (inct.splink.org.br). Managed as part of SpeciesLink by CRIA (Centro de Referência em Informação Ambiental; Center for Environmental Information), data from 130 Brazilian herbaria and 20 foreign herbaria make digitized data from more than 7 million specimens available for searching. SpeciesLink, under the leadership of Dora Canhos, is one of the world's most innovative specimen data portals, offering a variety of tools for refining specimen data and for predicting species distribution.

On the night of 2 September 2018, a fire devastated the main building of Rio's National Museum, the largest natural history museum in Latin America, resulting in the loss of more than 90% of the 20 million objects stored there. The loss to Brazil's scientific heritage is incalculable, made all the more painful because it highlighted the neglect of this priceless resource: the museum did not have a fire suppression system, and water hydrants closest to the museum were not functional. Thankfully, the 600,000 herbarium specimens held by the museum, stored in a separate building, were not damaged in the fire.

The People's Republic of China

A great diversity of plants and fungi flourish in China's tropical, temperate, and boreal habitats. In the warm temperate habitats of the country, remnants of the forest that encircled the northern hemisphere 15,000 years ago are larger than anywhere else in the world, meaning that many more of our planet's once-widespread species have persisted there. More than half of Chinese plants are endemic species, found nowhere else. China has more than 33,000 plant species within its borders, more than the United States and Canada combined, and the country has a long tradition of using them in herbal medicine, dating back about 5,000 years, approximately the same time that written language developed there. Over the millennia, the Chinese have accumulated a great deal of empirical knowledge about plants; some 8,000 Chinese plants have a medicinal or other economic use attributed to them.

Early Exploration

When Europeans first began to visit China in numbers some four centuries ago, they encountered a culture with a much longer history of plant knowledge than their own, established in the context of complex political and social institutions. The British East India Company, founded in 1600, was generally limited to China's southern coast, where it established a flourishing base in what is now Guangzhou. The plants the Europeans saw there attracted attention and foreign botanists. Scottish surgeon and naturalist James Cunningham (fl. 1698–1709) was the first European to send a botanical collection home from China; it consisted of 600 specimens (now on deposit at the Sloane Herbarium) and seeds for cultivation, among them seeds for *Cunninghamia lanceolata* (Chinese fir), a widely grown ornamental tree.

Frustrated by lack of access to China's interior and north coast, the East India Company sent troops to the country in 1839 to force the Chinese to buy supplies of opium from them to raise funds for their operations. The so-called Opium War resulted in cash the British could use to establish profitable tea plantations in India and Sri Lanka; it also gave them the right to establish many more trading stations in China. Botanically, the Qing Dynasty's defeat opened up a great deal of China's interior for plant exploration, and one of those who took full advantage of this opportunity was Augustine Henry (1857–1930). Born in Scotland, raised in Ireland, and trained as a physician, he entered the Imperial Customs Service in China in 1881 and the following year was sent to the remote posting of Yichang in Hubei province to investigate plants used in Chinese medicine. Unlike most previous explorers, Henry became proficient in Chinese, a skill that greatly facilitated his travels. By 1896 he had collected 15,000 specimens of 5,000 species, 500 of them new to science. He also recorded the native names and applications for plants used in traditional Chinese medicine. Henry's early collections went to Kew's herbarium, his later ones to various other herbaria. The herbarium of the National Botanic Gardens in Glasnevin, Ireland, is named for him, as are many plants, including the widely cultivated *Clematis henryi*.

Clematis henryi, a favorite garden clematis, has some of the largest white flowers in the genus.

When Charles Sprague Sargent (1841–1927) began to develop the Arnold Arboretum in the Jamaica Plain district of Boston, Massachusetts, he saw China as a possible source of new ornamental plants for American gardens. After ascertaining that many Chinese plants would grow well in the area, Sargent made the acquisition of more of them a top priority, with a preference given to trees and shrubs. He hired a series of men to collect plants throughout China, instructing them to bring back herbarium specimens as well as plants and seeds for cultivation. Sargent's collectors included Joseph Rock, Frank Kingdon-Ward, and perhaps the most successful, Ernest Henry Wilson, who succeeded him as director of the Arnold Arboretum.

Founding Botanists

The last emperor abdicated in 1912, leading to the establishment of the Republic of China. The new republic was anxious to join the community of nations and was notably more outward-looking. One of its aims was the incorporation of Western

ISOLECTOTYPE OF:

Clematis henryi Oliv.
in Hook. Icon. tab. 1819 (1889)

Aleck Yang   20-I-1995

NY IMAGED

DR. AUG. HENRY'S COLLECTIONS FROM
CENTRAL CHINA, 1885-88.

NO. ~~~~~~~~~~

= 266, 3280

Clematis Henryi, Oliv. sp. n.

Prov. HUPEH.

▲ Arnold Arboretum dendrologist John George Jack (left), examining maple leaves at the arboretum in the summer of 1917. The student on the right is thought to be Chen Huan-Yong.

◀ A set of Henry's specimens, including this of *Clematis henryi*, was purchased by the New York Botanical Garden in 1901 using funds donated by garden members.

science into its educational systems and Chinese life. Universities were founded in China starting in the 1890s, many of them helped by the United States or other foreign nations. For advanced study, Chinese students up to the time of World War II generally traveled to the United States or Europe. The Boxer Rebellion, of Chinese against imperialist foreign powers, was quelled in 1901, and China was compelled to pay indemnity funds to eight foreign powers for the deaths and damages that resulted from the uprising. Alone among these, the United States used a large part of the funds they received to assist Chinese students in coming to America for advanced study, thus helping greatly in the development of science and other academic fields in the country. All four of the Chinese botanists profiled here received training in the United States. Returning home, they each made pioneering contributions to the documentation of Chinese biodiversity and to the training of subsequent generations of Chinese botanists.

Chen Huan-Yong (1890–1971) was born in Hong Kong, the son of a Chinese diplomat and his Cuban wife. He went to Harvard in 1915 specifically to study the rich collections of Chinese plants at the Arnold Arboretum. When he returned home, a fellowship from Harvard allowed him to continue to document China's plants over the following years. He focused his collecting first in Hubei province, where a retired collector who had worked with Augustine Henry assisted him. His specimens were deposited at Nanjing University, where he began his teaching career, and he shared duplicate specimens with Harvard. Chen later moved to Sun Yat-sen University in Guangzhou, establishing the herbarium there. In 1922 he published *Chinese Economic Trees*, a work begun while he was at Harvard, and through the 1930s he maintained an active field program. By the end of the decade he was the leading figure in Chinese botany.

Hu Xiansu (1894–1968) was born in Nanchang, Jiangxi province. He obtained a bachelor's degree from UC Berkeley in 1916 and went on to become the first Chinese student to obtain a doctorate from Harvard, carrying out his studies at the Arnold Arboretum. He co-founded the Fan Memorial Institute of Biology in Beijing in 1928, becoming its second director in 1932. There he built the institute's herbarium into the largest in China, amassing more than 185,000 specimens in a decade. Like

Chen, he started with a program of exploration, organizing expeditions to Zhejiang, Jiangxi, and Fujian provinces; these collections formed an important nucleus for the institute's herbarium. Hu also sent specimens on exchange to the herbaria at the Arnold Arboretum and the Berlin Botanical Garden, thus providing (and obtaining) comparative material for the study of the Chinese flora. Together with Chen, Hu published *Icones Plantarum Sinicarum* in five large-format volumes between 1927 and 1938. The series included descriptions in Chinese and English as well as life-sized drawings of plants.

Marine phycologist Zeng Chengkui (1909–2005) was raised in Xiamen, an island city on China's coast, where there was a thousand-year-old tradition of cultivating algae for human consumption and using algae as animal feed and fertilizer. This history inspired Zeng to devote himself to broadening the traditional practice of algal culture into a commercially viable business that would support local farmers. The first step toward realizing his ambition was to learn to identify algae. He began to

◀ Specimen of the palm *Phoenix loureiroi*, collected by Chen in 1929 and obtained by the New York Botanical Garden from Sun Yat-sen University.

▼ Hu Xiansu and fellow members of the Academia Sinica, 1948.

Hu Xiansu

collect and preserve specimens, sending those he could not name to specialists in Japan and the United States. He received a fellowship from University of Michigan to study algal taxonomy with William Randolph Taylor, a leading expert, arriving there in 1940. Zeng's dissertation was a treatment of the red algae of Hong Kong.

Mycologist Deng Shuqun (1902–1970) was born in Fuzhou, Fujian province. After graduating from Beijing's Tsinghua University in 1923, he received a fellowship to study plant pathology and forestry at Cornell with mycologists Herbert H. Whetzel and Harry M. Fitzpatrick. He obtained a master's degree at Cornell and then completed coursework for a doctoral degree but not his dissertation, because in 1928 he accepted a professorship at Lingnan University in Guangzhou. Several years later he moved to Nanjing, where he held positions at several institutions, finally settling in 1933 at the Metropolitan Museum of Natural History, soon to become the National Institute of Zoology and Botany. The major work in Deng's early career was his English-language *Higher Fungi of China* (1939), which contained treatments of 1,400 species. During this period, Deng continued to build his own herbarium, as well as to send duplicate specimens to Cornell, where they were accessioned into the Plant Pathology Herbarium.

NYBG 775700

NYBG 775700

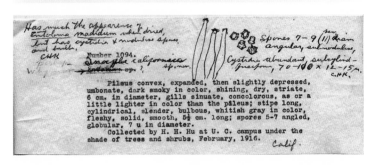

During World War II, the Japanese first occupied government centers in Beijing and Nanjing, then spread throughout the countryside. Academic work stopped as researchers and students fled the destruction that ensued. As they moved toward the southeast, Chinese scholars came into contact with unfamiliar plants and animals. One outstanding example was Hu Xiansu's discovery of living plants of the genus *Metasequoia*, which had been named in 1941 by a Japanese botanist for a fossil

◀ Type specimen (top view, underside, and notes) of *Inocybe californica*, collected by Hu in 1916 on the UC Berkeley campus; botanist and mycologist Calvin H. Kauffman at the University of Michigan recognized and described it as a new species.

▶ *Sargassum aquifolium*, collected in Hainan by Zeng Chengkui (as a variety of *S. echinocarpum*) in 1934.

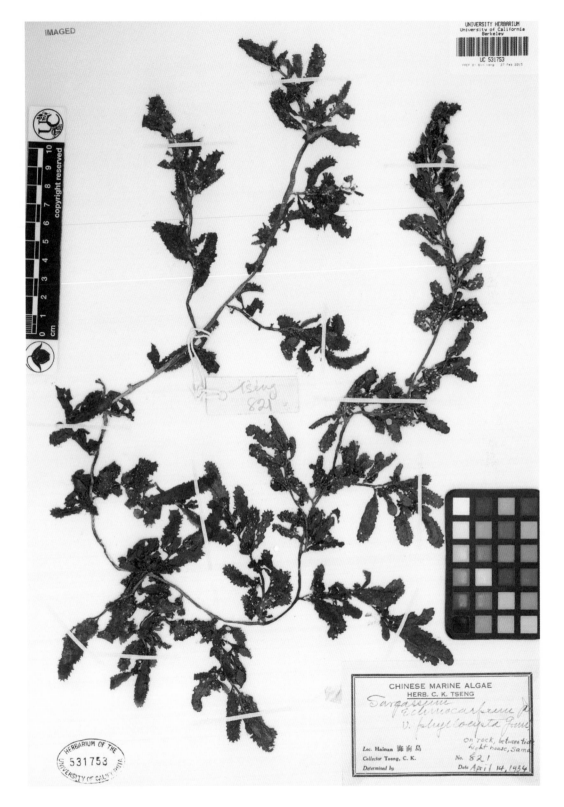

plant known widely across the northern hemisphere. When living plants of an unusual conifer were discovered in Sichuan and Hubei provinces the same year, Hu and colleague Zheng Wanjun determined that it belonged to the same genus as the fossil. They described the extant plant as *Metasequoia glyptostroboides* (dawn redwood). In 1948 the Arnold Arboretum led an expedition to collect seeds of the plant, which were distributed to various botanic gardens around the world.

Deng Shuqun, pictured here with Julian H. Miller, a fellow graduate student in mycology, and Arthur H. Chivers, professor of botany at Dartmouth College, at Taughannock Gorge in upstate New York, 1926.

Zeng Chengkui was in the United States when the war began, unable to return home. He spent the war years working at the Scripps Institution of Oceanography in La Jolla, California, where he was employed in a government-funded war effort program to find a domestic source of agar, a product derived from red algae. Agar has a wide range of uses in foods and biomedical research. By the end of the war he was a leading expert on growing and processing marine algae, writing technical and popular articles and giving invited lectures. He returned home to a position at National University of Shandong in Qingdao, planning to found China's first ocean research institute and pursue his dream of cultivating algae for food and commercial applications. He was unprepared for how difficult postwar conditions would be in Qingdao but eventually established the Institute of Oceanology there and took steps toward developing a mariculture program.

Deng Shuqun moved to Chongqing in Sichuan province at the beginning of the war, when the National Institute of Zoology and Botany relocated there. He took with him about 2,000 of his fungal specimens. Even in Chongqing, hundreds of miles from the center of the conflict, Deng was worried about the safety of his specimens. So, he divided the specimens in his herbarium into two sets, keeping one set for himself (it did not survive the war) and sending the other to the United States. The specimens destined for America were carried by ox cart to Indochina and then by ship to Washington, D.C. There the specimens were divided once more—one set was deposited at the U.S. National Fungus Collections in Beltsville, Maryland, and

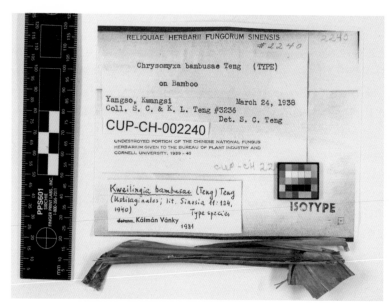

▶ Type specimen of the parasitic rust fungus *Kweilingia bambusae*, collected (as *Chrysomyxa bambusae*) by Deng in 1934, from Cornell's Plant Pathology Herbarium.

▶▶ A 49-million-year-old foliage spray of *Metasequoia occidentalis* from Washington State's Klondike Mountain Formation.

the other was sent to Cornell's Plant Pathology Herbarium. Deng's specimens have been digitized and are available online through the Mycology Collections data portal (mycoportal.org), which contains digital records for most fungal collections housed in American herbaria. In 2009, Cornell University divided many of their duplicates of the Deng fungi in order to return a set to China.

Modern Documentation and Herbaria

After World War II, national government forces and the Communist Party of China gained control of the government, and science in China was completely reorganized. The Academia Sinica, founded in 1928, was split in 1949 into a mainland-based Chinese Academy of Sciences (CAS), while the name "Academia Sinica" continued for the analogous body in Taiwan. On the mainland, the CAS became the umbrella organization for Chinese scientific research institutes, following the pattern that had been developed in the Soviet Union, and university-based research was deemphasized. Scientific work was valued in the early years of the People's Republic, prized not only as a component of China's recovery from the war but also as an important part of the country's modernization effort.

Zeng Chengkui's work on mariculture was particularly encouraged, and he began an alga farming program using a brown alga (*Laminaria* sp.); his program was touted as a success of China's Great Leap Forward in 1958. Deng Shuqun continued his documentation of Chinese fungi during this time; his *Fungi of China* (1963), a monumental Chinese-language work, remained the classic text for Chinese mycologists for many years. Chen Huan-Yong's institute at Sun Yat-sen, brought under auspices of the Chinese Academy of Sciences in 1954, was renamed the South China Institute of Botany. That same year, Chen was among the scientists who attended the First National People's Congress, presided over by Mao Zedong, in which Mao urged participants to learn from the advanced experience of the Soviet Union. Chen continued to publish the results of his botanical research; before 1950 he always published in English, afterward only in Chinese.

An outspoken person on political and economic issues, Hu Xiansu did not fare as well under the PRC government. His position was downgraded after the Fan Memorial Institute came under the control of the Academy of Sciences. He continued his research, however, focusing on economic botany and textbooks, and continued to translate foreign books and papers into Chinese. He published a key to families and genera of vascular plants of China (1953–54) that summarized much of his life's work.

The Great Proletarian Cultural Revolution began in 1966, bringing to a halt most scientific research. Our four Chinese scientists fared badly during the period. Hu, in poor health before the Cultural Revolution, was labeled a reactionary academic authority, criticized for his work with foreigners, fired from his job, and isolated. He died of a heart attack in 1968. Chen and Deng similarly did not survive to see the end of this harsh chapter in modern Chinese history. Their deaths received no notice at the time, but all were later rehabilitated, and Hu's ashes were interred at the Lushan Botanical Garden in Jiangxi province, which he founded.

Zeng was not only labeled but imprisoned as a reactionary academic authority—starved, beaten, forced to write confessions. By 1970, however, he was once again allowed to participate in applied research at the Institute of Oceanology. His status improved even more after the historic meeting of Mao and Zhou Enlai with Richard Nixon to reestablish diplomatic relations in 1972. Selected as a member of a delegation of Chinese scientists to the United States in 1975, Zeng revisited the Scripps Institution, and when the delegation was invited to the White House, he met President Gerald Ford, a fellow University of Michigan graduate. In 1978 Zeng was appointed director of the Institute of Oceanology and started an ambitious program to revitalize Chinese marine science and technology.

Following détente, communication between Chinese and western botanists was reestablished, resuming officially in 1978 with a visit from a delegation of U.S. botanists representing the Botanical Society of America. The 1980 Sino-American Botanical Expedition to the Shennongjia Forest District and the *Metasequoia* region of Lichuan (both in Hubei province) was the first botanical collecting trip by American scientists to China since 1949. In 1981, Chinese botanists attended the International Botanical Congress in Sydney, Australia, where many were able to meet

Zeng and Gerald Ford, both University of Michigan graduates, at the White House.

their international colleagues for the first time.

The idea for a flora of China emerged in 1933 with the founding of the Botanical Society of China, but work did not begin on the project until 1959. Some volumes of the *Flora Reipublicae Popularis Sinicae* (Flora of the People's Republic of China) appeared before and (remarkably) even during the Cultural Revolution. Work on the flora resumed in earnest in the early 1970s and ramped up during the next two decades as government funding provided financial support to increase production rate, with the final volume completed in 2004. Some 320 scholars from 71 institutions—working largely without benefit of international collaboration or information exchange with the outside world—contributed to this 80-volume work, considered one of China's key research projects of the 20th century.

A collaborative project (among four Chinese and seven non-Chinese institutes) to produce an English-language *Flora of China* began in 1988; the Institute of Botany in Beijing and the Missouri Botanical Garden served as coordinating centers for the project. The first volume was published in 1994 and the final in 2013. All treatments are co-authored by Chinese and non-Chinese authors. In addition to printed volumes, a version of the flora is available online at efloras.org.

As for herbaria, the first Chinese-managed institutional herbarium was founded in 1915 at the Class One Agricultural School in Nanjing, Jiangsu province, later Nanjing University. There, both Hu and Chen began their teaching careers and received their first experience in herbarium management. They would later use this knowledge in the founding of herbaria that are today among China's largest: the National Herbarium in Beijing, which incorporates the Fan Memorial Institute collection, and the South China Botanical Garden's herbarium, which incorporates Chen's collections from Sun Yat-sen University.

Herbaria, like most Chinese institutions, suffered greatly during World War II. Approximately half of the Fan Memorial Institute's specimens were either destroyed or taken to Japan, never to be recovered. Arnold Arboretum director Richard Howard, who was part of the 1978 delegation to China, reported that generally Chinese

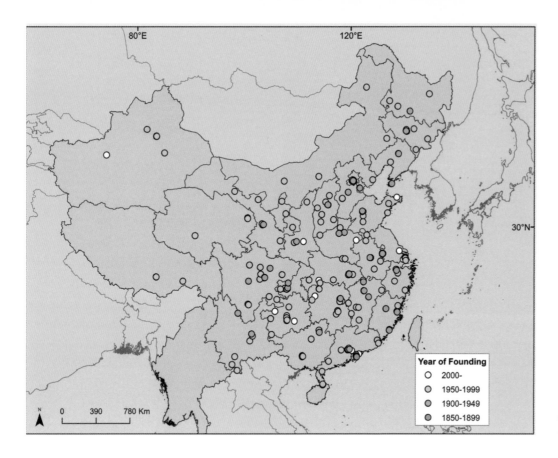

Map showing location of Chinese herbaria, points color-coded by year of founding.

herbaria seemed adequately housed, with room for expansion, although specimens were kept in wooden cases and thus required extensive chemical treatment to prevent insect predation. He further noted that specimen labels generally contained only a small amount of information and that specimens were mounted on low-quality paper. Most institutions had considerable backlogs of unprocessed specimens.

At the time of Howard's visit, China had 216 herbaria; today, there are 361, containing a total of more than 20 million specimens. Half of the herbaria are associated with universities, 37% with government-sponsored research institutions, 7% with botanic gardens, parks, and nature reserves, and the balance with museums.

Chinese herbaria are engaged in digitizing their specimens. The Chinese Virtual Herbarium (cvh.ac.cn/en), the first online portal for herbarium specimen holdings, is now embedded within the National Specimen Information Infrastructure (NSII), a digital infrastructure for sharing data from Chinese collections of plants, animals, fossils, and minerals. Developed to support national initiatives in biodiversity conservation, this resource contains more than 15 million specimen records, 6 million images, and digital versions of 18,000 botanical books and articles. Besides holding

Botanical sculpture in front of the Shenzhen Convention and Exhibition Center, site of the 2017 International Botanical Congress.

the index of scientific names for all Chinese plants, the NSII is a comprehensive reference for the various vegetation types found in China, complete with images of the species that make up each type. The NSII also provides content geared for educational use (primary through university level) and the public.

A crowning accomplishment of the Chinese botanical community was the 19th International Botanical Congress, hosted in Shenzhen, Guangdong province, in 2017. Approximately 7,000 botanists from around the world attended this conference. Six days were devoted to a variety of topics in current botanical research, with the need for the greening of cities as an overarching theme. The pioneers who worked so hard to bring western approaches to plant study to China would have been gratified by the leadership shown by their homeland in the field of study to which they devoted their lives.

South Africa

South Africa contains only 2% of the world's land surface but is home to approximately 8% of all plant species. It contains a variety of distinct biomes that contrast sharply with one another. Take the Fynbos, for example, an area of mostly shrubby plants, many with spectacular flowers: of the approximately 9,000 species that grow there, in the region around the Cape of Good Hope, 68% are endemic to the region. Likewise the Succulent Karoo biome, north and east of Cape Town and extending into Namibia, contains some 3,700 plant species, almost half of which grow nowhere else.

Fynbos vegetation south of Cape Town.

Early Exploration

South Africa was home to indigenous people speaking at least 35 languages when Europeans first colonized the region. Even before the Cape Colony was established, voyagers who stopped in the region noted its distinctive plants and took some back to Europe for cultivation; an illustration of the flowerhead of a protea, one of the most iconic Fynbos plants, appeared in Charles de L'Écluse's *Exoticorum Libri Decem* (1605). In 1652, under the leadership of Jan van Riebeeck, the Dutch established the first settlement in the region for the purpose of resupplying ships of the Dutch East India Company on their journeys between Europe and the East Indies. Van Riebeeck brought gardener Hendrik Boom with him from Amsterdam, and on 29 April 1652,

Succulent Karoo vegetation featuring *Pachypodium namaquanum*, which, although it resembles a cactus, is actually in the milkweed family.

Boom broke ground in what would become the first garden in what is now Cape Town, to grow fresh food for passing ships and cultivate indigenous plants that might become a source of income for the colony.

Linnaean student Carl Thunberg spent three years in what is now South Africa, arriving in Cape Town in 1772. He made three journeys into the interior, collecting some 3,100 specimens that are maintained as part of the herbarium at Uppsala University. His publications on the South African flora, *Prodromus Plantarum Capensium* (1794, 1800) and *Flora Capensis* (1807–13), are foundational works for the study of South African plants.

Many other European collectors visited South Africa in the late 18th and early 19th centuries. William J. Burchell (1781–1863), an English explorer and naturalist, is among the most notable. He collected some 50,000 specimens, the first set of which are on deposit at the Kew herbarium. His South African collections contained more than 2,000 new species, some described by Burchell himself. He captured his experiences in the two-volume *Travels in the Interior of Southern Africa* (1822, 1824). Burchell's exploration and publications factored into the British government's decision to send emigrants in 1820 to settle in the Eastern Cape, establishing the district of Albany.

German horticulturalist and botanical explorer Johann Franz Drège (1794–1881) collected widely in the Western and Eastern Cape provinces and KwaZulu-Natal between 1826 and 1833, making over 200,000 well-documented collections of more than 8,000 species. He gave a set of his specimens to his friend Ernst H. F. Meyer, who included some of them in his *Commentariorum de Plantis Africae Australioris* (1835–37). Part of the remainder of Drège's collections were lost to fire in Hamburg in 1842, but duplicates of his collections are widely distributed in other herbaria.

Founding Botanists

Even before unification in 1910, when South Africa officially became the country's name, some early Cape Town institutions, such as the South African Museum (now Iziko South African Museum), founded in 1825, flirted with this designation. The region actually consisted of four colonies. Two of these, the Cape and Natal (now KwaZulu-Natal), were British; the others, the Transvaal and the Orange Free State, were independent Boer republics. The Boers were European settlers, mostly Dutch and German; the majority of them settled in the interior regions of South Africa, where they developed a distinctive language, Afrikaans, derived from Dutch. Botanical infrastructure developed separately in the British and Boer colonies up to the time of unification and somewhat beyond. Cape Town was the largest city, the most advanced culturally and educationally, and the most European in outlook for most of the 19th century. Botanical activity there was strongly tied to the Royal Botanic Gardens, Kew. The Boer republics lacked a close relationship with European botanical institutions and early on had fewer interactions with the international botanical community.

▲▲ The Company's Garden, originally developed by Hendrik Boom, is preserved as a city park in Cape Town.

▲ *Protea cynaroides*, the national flower of South Africa and namesake of the country's national cricket team, the Proteas.

"The Rock Fountain, in
the Country of the Bush-
men," a plate from Burchell's
*Travels in the Interior of
Southern Africa*.

The first official colonial botanist in the Cape Colony was Karl Wilhelm Lud-
wig Pappe (1803–1862), a German-trained physician with an interest in systematic
botany, forestry, agriculture, and plant disease. He was a regular correspondent
of William Hooker at Kew. Pappe's specimens were studied for a second *Flora
Capensis*, by English botanist William H. Harvey and his German colleague Otto
W. Sonder, published in three volumes plus a supplement between 1860 and 1865.
Pappe's collections formed the basis of the first herbarium in the region, at the South
African Museum (now part of the Compton Herbarium).

Harry Bolus (1834–1911) was a Cape Town stockbroker who became an enthusiastic plant collector, specializing in orchids. He amassed what would become one
of most important plant collections in the country and founded its first academic
position in botany, at South African College (now the University of Cape Town).
The Bolus Herbarium was incorporated there upon his death and remains the largest university herbarium in South Africa. Among its holdings are the collections of

Elsie Esterhuysen (1912–2006), a Cape Town native and one of the most outstanding collectors of the South African flora. She collected 36,000 specimens during her career, among them some 150 new taxa. Two genera and 56 species are named for her. Louisa Bolus, daughter-in-law to Harry, was Esterhuysen's equally accomplished colleague at the herbarium and another of South Africa's great female botanists; no other woman has named a larger number of plant species, 1,494 in total over the years (although some have since been synonymized).

The first occupant of Harry Bolus' botany chair at South African College was Henry H. W. Pearson (1870–1916), a Cambridge-educated British emigrant. He founded the Kirstenbosch National Botanical Garden at the base of Table Mountain, considered by many to be one of the world's most beautiful botanic gardens. Kirstenbosch did not have an herbarium until the collections of the South African Museum and the Government Herbarium, Stellenbosch, were brought together there (in 1957 and 1996, respectively) and incorporated with the existing Compton Herbarium, founded in 1937; the amalgamated Compton Herbarium, which remains the second-largest in the country, is named for Robert H. Compton, who succeeded Pearson as director of the garden in 1919.

Peter MacOwan (1830–1909) was a British chemist who emigrated to South Africa seeking a better climate for a lung ailment. His health greatly improved after the move, and he developed an interest in the plants and fungi of his new home, collecting widely from his home base in Grahamstown (now Makhanda) in

Kirstenbosch National Botanical Garden near Cape Town, South Africa.

Specimen of *Mesembryanthemum noctiflorum* subsp. *stramineum*, described by Louisa Bolus (as *Aridaria esterhuyseniae*): the Mesembryanthemaceae (now Aizoaceae, a family of popular succulents) was her specialty.

the Eastern Cape province. He corresponded and exchanged plant specimens with Joseph Hooker at Kew and Asa Gray at Harvard. He sent his fungal collections, which included macrofungi as well as plant pathogens, to European mycologists Mordecai C. Cooke and Károly Kalchbrenner of Budapest, and together these two published, in 1880, the first scientific article on the fungi of South Africa, describing 120 species. MacOwan later became a consultant to the Cape government on plant diseases and served as curator of their herbarium. To help meet the demands for South African plant and fungal specimens, MacOwan formed the South African Botanical Exchange Society, a group of amateur botanists who collectively distributed some 9,000 duplicate specimens of South African plants to herbaria in Europe and North America, receiving specimens from these areas in return. Although he spent most of his effort on botany after settling in the Eastern Cape, he did use his chemistry background there for an important purpose: MacOwan was part of the team who verified that a transparent rock found by a boy on his father's Northern Cape farm near Hopetown in 1867 was indeed a diamond, thus beginning an industry with profound economic and social implications for South Africa and beyond. MacOwan's collections are widely distributed, but most are preserved in the Compton and Albany Museum herbaria (plants) and the herbarium of the Plant Protection Research Institute in Pretoria (fungi).

The Durban Botanic Gardens was established in 1849, and a colonial herbarium (now the KwaZulu-Natal Herbarium) was established there in 1882 by John Medley Wood (1827–1915). Wood, a self-trained botanist, collected all groups of plants and fungi, but he specialized in ferns. He sent his bryophyte collections to Polish botanist Anton Rehmann, who distributed them in *Musci Austro-Africani*, an exsiccata consisting of approximately 700 numbers of South African specimens. Olive Mary Hilliard (b.1925), a prolific collector of plants in the Drakensberg region of the province, was named curator of the KwaZulu-Natal Herbarium in 1963. She has authored 372 plant names, the fifth-highest total by a woman.

As the interior became more accessible in the mid-1800s, the pace of botanical collecting there increased; and with the discovery of gold in the Transvaal, the balance of power and prestige changed among the Boer colonies, leading to the founding of Johannesburg and Pretoria and the development of botanical institutions in those cities. The first official botanist of the Transvaal, and its first female civil servant, was Reino Leendertz (1869–1965), a Dutch emigrant. In appointing Leendertz to her position as assistant botanist in 1898, the director of the State Museum (later the Transvaal Museum) noted that although it was contrary to the policy of the committee to appoint a woman to the scientific staff, her exceptional qualification had led him to waive this objection. Aside from a brief interval during the

▶ The cycad *Encephalartos woodii*, named for John Medley Wood, is now extinct in the wild; all specimens in cultivation, such as this one in the Temperate House at Kew, are clones of the plant that yielded the type.

◀ South Africa's National Herbarium is on the grounds of the Pretoria National Botanical Garden.

Anglo-Boer War when she returned to the Netherlands, Leendertz spent her entire career at the museum. She collected more than 6,000 specimens for the herbarium she established there and co-authored a "first checklist" of 3,200 flowering plant and fern species occurring in the Transvaal and Swaziland, published in 1912. Her collaborator, Joseph Burtt Davy (1870–1940), was a British agriculturalist trained at the University of California who had immigrated to South Africa for a position in the Transvaal Department of Agriculture. He immediately began documenting the native flora of his new homeland, establishing the Transvaal Colonial Herbarium within the department. The herbaria built by Leendertz and Burtt Davy eventually became part of what is now the National Herbarium in Pretoria, established in 1903.

The designation of the Pretoria collection as the national herbarium was controversial—the botanists associated with the national botanic garden at Kirstenbosch believed that the national herbarium should be located on their grounds. However, the botanists of the Transvaal had a powerful ally in the choice of Pretoria for this honor. Jan Smuts, a military leader and statesman, was also a keen botanical collector. Although born in the Cape Colony, Smuts fought alongside the Boers and afterward lived in the Transvaal. He collected plants actively from 1924 to 1939, between his two terms as prime minister of the Union of South Africa, contributing to several accounts of southern African vegetation that appeared in the *Kew Bulletin* in the 1930s. Smuts is more often remembered as a vocal supporter of the segregation of the races in South Africa, helping to set the stage for apartheid, although in later life he retreated somewhat from these views. His specimens are housed at the

M.S.del. J.N.Fitch lith.

Vincent.Brooks,Day&Son Ltᵈ imp

Stapelia leendertziae (black bells) from *Curtis's Botanical Magazine* (1914), vol. 140, plate 8561. This species, named in Reino Leendertz's honor, is a popular container plant.

National Herbarium in Pretoria, and he is eponymized in several epithets; one of them, *Pteronia smutsii*, was named by Kew botanist John Hutchinson, who made several expeditions with Smuts.

Modern Documentation and Herbaria

For many years after unification, the Cape Town and Transvaal spheres of botany remained distinct. They eventually merged, however, and by the time the Republic of South Africa was formed in 1961, the development of a network of national botanic gardens across the country was well under way. Establishment of this network marked the beginning of a more unified approach to documenting and preserving the flora of the nation. Although South Africa's national botanic gardens originally focused on local floristic research, a 1984 government mandate reoriented their scientific focus toward plant conservation and broadened their scope to include the flora of the entire southern African subcontinent, using information obtained from as many fields of research—anatomy, genetics, physiology, chemistry—as possible.

In order to manage the data needed for a more holistic taxonomic understanding of South African plants, the National Herbarium initiated PRECIS (PREtoria Computerized Information System) in the early 1970s. This was the first major electronic database of plant information in the world, but since it predated the Internet, the resource was only available locally. The initial goal of PRECIS was to catalog the collections of the National Herbarium as the foundation for its major product, BODATSA, a checklist of all southern African plants. This resource has been maintained and updated ever since that time, providing a unified set of accepted plant names for subsequent floristic work across the country.

The formation in 1989 of the National Botanical Institute, headquartered in Cape Town, amalgamated the national gardens and the botanical research communities. In 2004 it expanded into the South African National Biodiversity Institute (SANBI), headquartered in Pretoria, with objectives of assessing the botanical research and conservation needs of the country; addressing those needs through development and maintenance of collections of specimens and plants living in national botanic gardens and herbaria; preserving endangered species; exploring the economic potential of native plants; and promoting understanding and appreciation of plants by citizens. Among SANBI's activities is the creation of an electronic flora of South Africa (sanbi.org/biodiversity/foundations/biosystematics-collections/e-flora) that will contain digitized versions of taxonomic descriptions of species occurring in South Africa derived from previous publications.

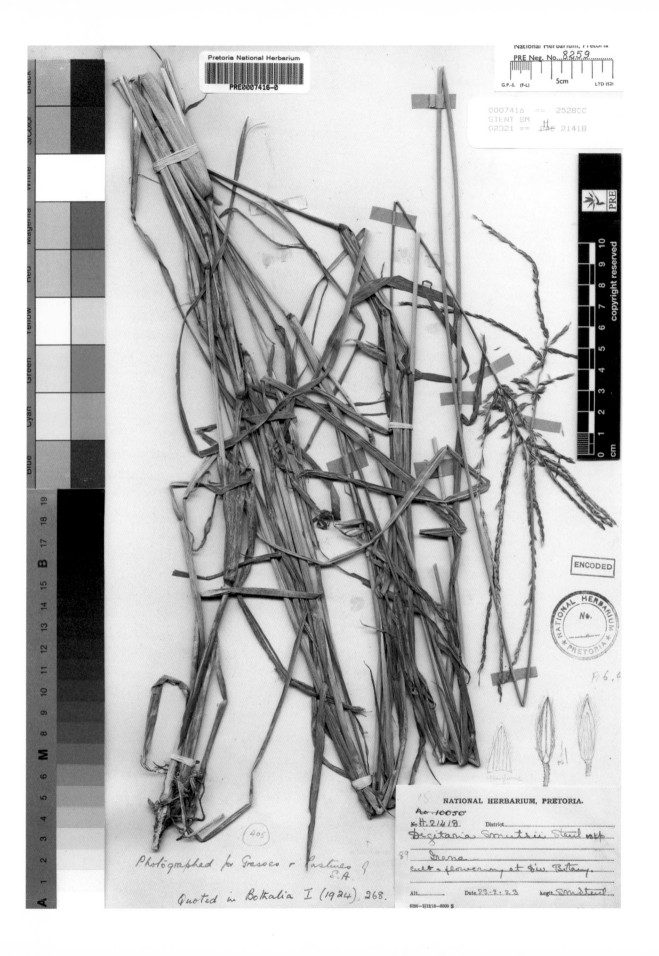

copyright reserved

PRE

ENCODED

NATIONAL HERBARIUM, PRETORIA.

No. 10050

No. H. 21418   District...........

Digitaria Smutsii Steud nov sp

89   Irene

cult. + flowering at div. Botany.

Alt............   Date 23.3.23   Legit. SM Stent

8320-3/12/18-5000 S

(405)

Photographed for Grasses + Pastures of
S.A.

Quoted in Bothalia I (1924), 268.

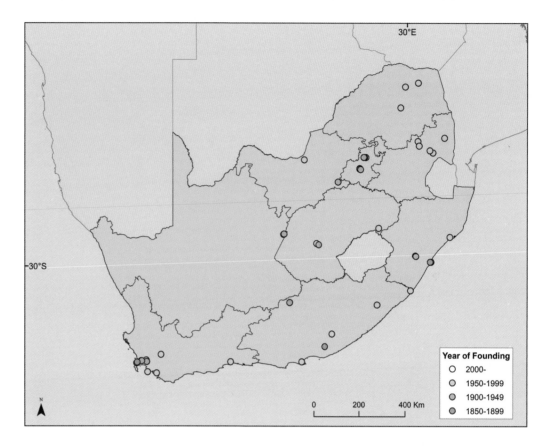

◄ Type specimen of *Digitaria eriantha,* a grass grown as a livestock forage crop in tropical and subtropical climates. The original species epithet was one of several that honored Jan Smuts.

▶ Map showing location of South African herbaria, points color-coded by year of founding.

Today approximately 3.2 million specimens are held in 46 herbaria in South Africa. The ten largest herbaria hold about 90% of all specimens. The smaller herbaria housed at nature reserves are usually focused on biodiversity within a small region or dedicated to particular kinds of plants, such as weeds or agricultural species. Most of the larger herbaria have digitized their specimens and share their data online through various portals, including the Global Biodiversity Information Facility. Only about 1% of the approximately 2,000 people who have collected plants in South Africa are indigenous Africans, but today about 15% of herbarium directors and staff are of native African heritage.

As the African country with the largest population of people of European heritage, it is not surprising that the herbarium tradition in South Africa is stronger than elsewhere on the continent. Yet, the 90 non–South African herbaria contain about 6 million specimens with 480 associated staff. Kenya has the strongest herbarium tradition of African herbaria next to South Africa, with 1.3 million specimens, followed by Egypt, with 733,000 specimens.

The Future *of* Herbaria

Thanks to centuries of work by explorers and scientists world-wide, there are now approximately 3,300 herbaria in 178 countries. Together they hold approximately 390 million specimens. If the locations of discrete collection events represented in this specimen total were spaced equally over the globe, there would be about five herbarium specimens for each square mile of the earth's land surface: one collected in the 19th century or earlier, one between 1900 and 1949, two between 1950 and 1999, and one since 2000.

Herbaria still serve their original function—to document the occurrence of plants and fungi and provide a reference for their identification and characterization. However, recent technological advances that allow us to study life at both the molecular level and on a global scale can be applied to herbarium specimens to help address some of the most critical problems we face today. New ways of sharing information allow herbaria to demonstrate the importance of plants and healthy ecosystems to an audience far beyond the scientific community.

◀ An herbarium specimen of
Linum lewisii (Lewis flax).

Herbaria and Global Change

Understanding the impact of global change on earth's organisms and ecosystems is a central focus of current biological research. Multiple lines of evidence indicate that human impact is changing earth's cycling of key components of our air, water, and soil, but documenting such change in a scientific manner is tricky because it requires measurements and observations before and after the change event. Plants are rooted in the ground, inevitably exposed to environmental conditions in the place where they grow. Therefore, herbarium specimens provide one of the very few tangible sources of data about the environments in which plants and fungi lived before industrialization and through each successive period of technological advance since that time. Among the world's herbaria we have specimens from a broad temporal and geographic range, and we can use them to look for evidence of change in different places and in different groups of organisms, in order to confirm that the patterns we find are widespread.

Twenty-first-century analytical equipment makes it possible to identify and quantify chemicals from tiny amounts of plant or fungal tissue from herbarium specimens, sometimes even very old ones. We can analyze these compounds to understand an organism's genetic makeup, pollutants in the soil where it grew, and even the composition of the air above it.

DNA Studies

In the 1990s scientists discovered that gene sequences could be recovered from the tissue of herbarium specimens. The challenges that had to be overcome in order to do this are that DNA degrades over time, and the drying of the specimen breaks the DNA strands. However, the quantity of DNA can be increased after extraction using PCR (polymerase chain reaction), an amplification technique; and long gene sequences can be reconstructed from shorter ones. Although a researcher would probably always find it easier to obtain DNA from living material, extraction from herbarium specimens may be the only option for species that are rare or extinct in the wild. Sampling from already-collected, authoritatively identified specimens may save months of work and thousands of dollars, even with the extra time needed for extraction from herbarium specimens. As the techniques for extracting DNA from dried specimens have improved, the amount of tissue needed for extraction has been reduced; usually just a small portion of tissue from one leaf on a specimen is sufficient. Although curators may not allow all specimens to be sampled, most herbaria will permit some judicious sampling of ample specimens for DNA extraction.

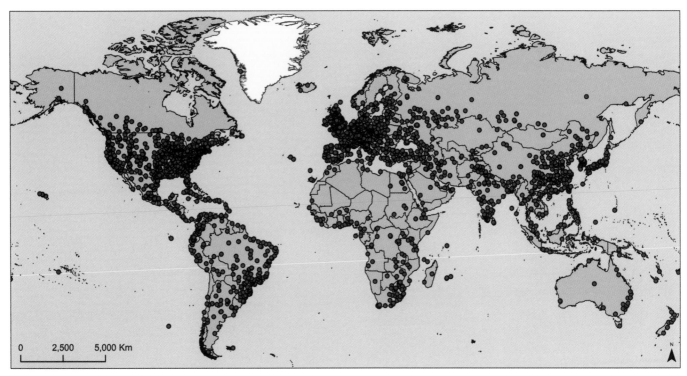

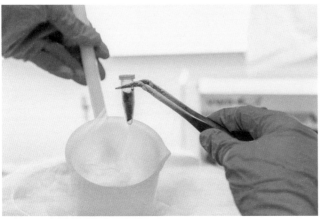

▲▲ Herbaria around the world.

▲ A sample of plant tissue being prepared for freezing in liquid nitrogen in the New York Botanical Garden's Pfizer Plant Research Laboratory. This is a preliminary step in the DNA extraction process.

Until recently, most DNA-based research using herbarium specimens contributed to delimiting species and constructing phylogenetic family trees to understand how organisms are related to one another. But the new and more powerful tools of next generation sequencing (NGS) now make it possible to sequence all the genes in an organism (its genome) and to detect the presence of any particular gene expression within it. Surveying the differences in genome structure among plants preserved as herbarium specimens allows researchers to understand which genes control traits such as height, leaf size, color, or flowering time, and how variation in the expression of these traits has developed in different locations and in different lineages of organisms. Such knowledge informs plant breeding programs to improve productivity in plants of economic importance, and to predict species tolerance to changes in environmental conditions. These techniques have also allowed herbarium specimens to be used to understand the spread of pathogens. Sequencing strains of *Phytophthora infestans* (the organism responsible for potato blight) from herbarium specimens enabled a reconstruction of the genetic changes that allowed the pathogen to destroy the potato crop in Ireland over a four-year period

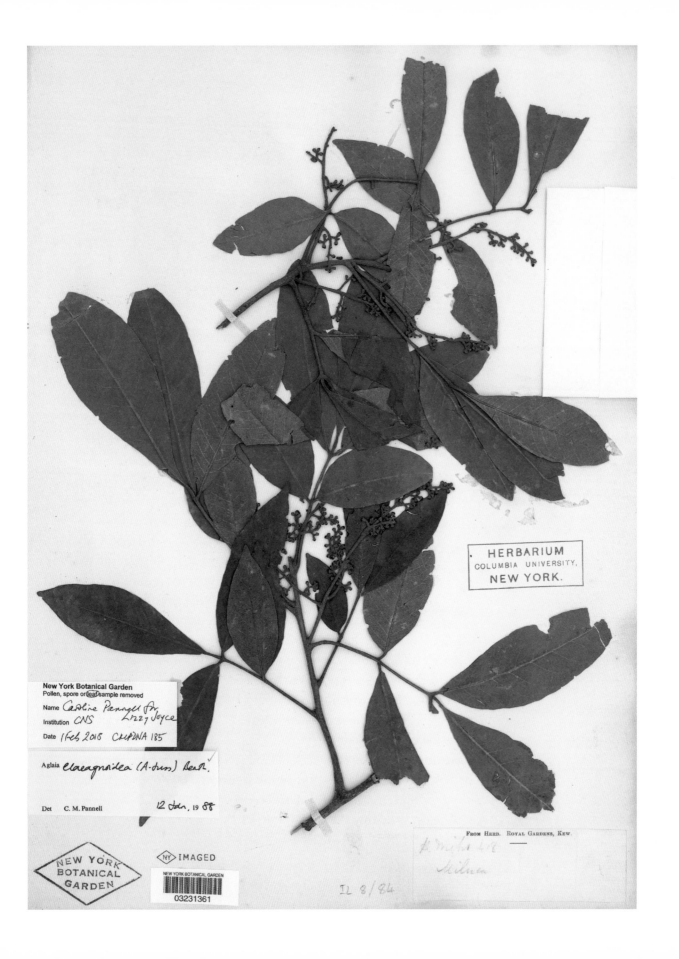

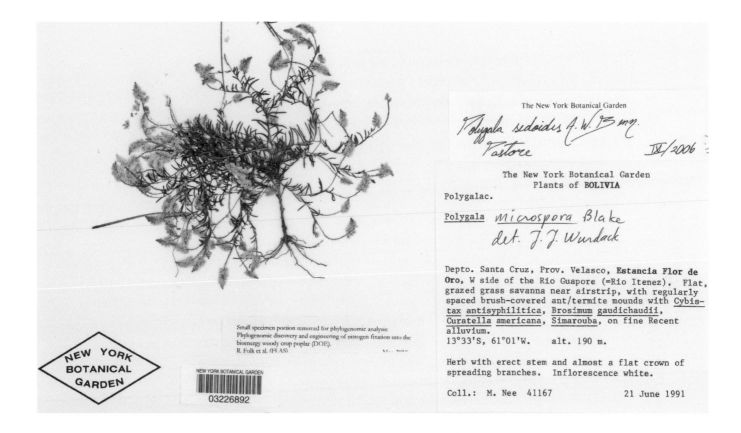

The New York Botanical Garden

Polygala sedoides A. W. Benn.
Pastore   IX/2006

The New York Botanical Garden
Plants of **BOLIVIA**
Polygalac.

<u>Polygala</u> *microspora Blake*
det. J. J. Wurdack

Depto. Santa Cruz, Prov. Velasco, **Estancia Flor de
Oro**, W side of the Rio Guapore (=Rio Itenez). Flat,
grazed grass savanna near airstrip, with regularly
spaced brush-covered ant/termite mounds with <u>Cybis-
tax antisyphilitica</u>, Brosimum gaudichaudii,
<u>Curatella</u> <u>americana</u>, <u>Simarouba</u>, on fine Recent
alluvium.
13°33'S, 61°01'W.   alt. 190 m.

Herb with erect stem and almost a flat crown of
spreading branches. Inflorescence white.

Coll.: M. Nee 41167   21 June 1991

Small specimen portion removed for phylogenomic analysis:
Phylogenomic discovery and engineering of nitrogen fixation into the
bioenergy woody crop poplar (DOE).
R. Folk et al. (FLAS)   Mar. 2010

NEW YORK BOTANICAL GARDEN
03226892

▲ *Polygala microspora*, just
one of the 10,000 herbarium
specimens imaged for NitFix,
a project led by Pamela and
Douglas Soltis and Robert
Guralnick of the University
of Florida, and sampled for
DNA. The goal of NitFix
is to engineer bioenergy
crops that can be grown in
marginal lands.

◄ The oldest specimen at
the New York Botanical
Garden from which DNA
has been successfully
extracted: *Aglaia elaeag-
noidea*, collected in India
by Robert Wight in 1835.

(1845–49), causing the Great Famine. Today an entire new field, museomics, focuses
on exploring the genetic diversity that can be recovered from herbarium and other
natural history specimens. Dried specimens are well adapted to this finer-grained
type of genetic analysis, because the first step in next generation sequencing is to
break the DNA strands into small pieces—the condition that already exists in DNA
from dried specimens.

Pollution Detection and Remediation

Iron, cobalt, and zinc are essential nutrients for people and plants, but other heavy
metals, such as cadmium, lead, and mercury, have adverse effects on us, even in tiny
quantities. These elements can reach harmful concentrations in our soil and drinking
water as a result of unregulated agriculture and industrial processes or inadequate
waste disposal methods. Generally less sensitive to these chemicals than animals,
plants and fungi absorb heavy metals in the water they take in from their roots or
hyphae, and small quantities of plant or fungal tissue from an herbarium specimen
can be sampled to determine the concentration of heavy metals where it grew.

Various studies have shown that heavy metal concentrations in herbarium specimens from industrialized regions are highest from times when manufacturing had few pollution control measures, especially the late 19th century in Europe and North America. A study of herbaceous plants in Rhode Island collected between 1846 and 2015 demonstrated that lead concentration in plant specimens decreased during this period, as manufacturing plants in the area adopted more effective emission controls; interestingly, lead levels in plants from Block Island (13 miles from the mainland), where there was no significant industry, were also high during the times of greatest pollution (Rudin et al. 2017). The decline of heavy metal concentrations in specimens collected after mitigation for contamination pollution can indicate the effectiveness of pollution control measures; for example, monitoring the lead composition in moss or lichen samples collected on roadsides in several areas of the world has demonstrated that removing lead from gasoline does indeed reduce the levels of this heavy metal in the vegetation and therefore also in soil and water. We can also use plant and fungal tissue to monitor the effect of documented pollution events, such as the release of radioactive isotopes from nuclear power plants. A study of lichen specimens from Macedonia in northern Greece indicates an increase in Cesium-137 for about two years after the Chernobyl reactor accident, followed by a gradual decline to pre-accident levels in subsequent years (Sawidis et al. 1997).

Documenting Past Atmospheric Conditions

Even before we had sophisticated techniques for molecular-level studies of plants, we learned that herbarium specimens can tell us about CO_2 levels in the atmosphere. Carbon dioxide enters the plant through stomates (from the Greek *stomata*, "mouth"), tiny holes on the underside of the leaf. Photosynthesis, which takes place in the interior leaf cells, uses energy from the sun to convert the CO_2 into sugars to fuel the plant's growth. Oxygen (O_2) is a product of photosynthesis, escaping the plant through the stomates and entering the lungs of animals to fuel our metabolic processes. Experiments in which plants were grown in controlled environments with different CO_2 concentrations in the air showed that the density of stomates on the underside of a plant leaf respond to these differences—the greater the CO_2 concentration in the atmosphere, the fewer the number of stomates needed to supply the plant with this key photosynthesis precursor. In one study, leaves from herbarium specimens of woody (and therefore long-lived) plants collected over a 200-year period were examined to see if the rise in atmospheric CO_2 that accompanied industrialization was reflected in the stomatal density; the study showed that indeed such

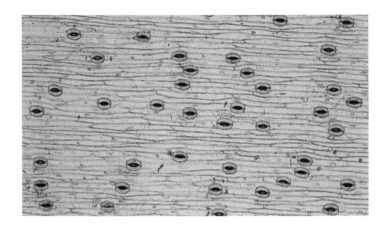

Microscopic view of the underside of a leaf showing the elongated epidermal cells and the kidney-shaped guard cells that surround the stomatal "black holes."

a correlation does exist (Woodward 1987). A variety of subsequent studies confirmed this result, including one that examined leaf material of *Olea europaea* (olive) from a broader time span, beginning with leaves from a wreath found in Tutankhamun's tomb (Beerling and Chaloner 1993).

Carbon dioxide is not strictly speaking a pollutant, since it occurs naturally in the air and plants require it to make food through the process of photosynthesis, but of course rising levels of CO_2 are implicated in the reduction of the ozone layer and warming of the earth. Also, rising CO_2 levels may make it more difficult for plants to obtain nutrients from the soil, which in turn lessens their food value for the animals that eat them. A study of nitrogen isotopes in plant leaves from herbarium specimens collected in Kansas grasslands between 1876 and 2008 demonstrated that concentrations have been declining over time, even with increased input of nitrogen through human activity (McLauchlan et al. 2010). These results provide support for the hypothesis that as atmospheric CO_2 levels rise, nitrogen is increasingly sequestered in ecosystems, that is, less of it is released to the soil through processes such as the decomposition of dead plant material.

Predicting Changes to Biodiversity

Most of the world's herbaria are now engaged in digitization on some scale and so far have collectively digitized and shared online about 81 million herbarium specimens of plants and fungi. These data are shared online through various portals, including the Global Biodiversity Information Facility (gbif.org), the Atlas of Living Australia (ala.org.au), and Brazil's SpeciesLink (splink.cria.org.br). The adherence of the data shared through these portals to a common standard, the Darwin Core, ensures that data sets from a variety of sources can be combined as needed to address a particular research question of interest.

If we are interested in whether a particular species is at risk for extinction, we can enter collection localities from digitized specimen data into a Geographical Information System (GIS) to determine its distribution range. This is the first step toward determining whether or not a species is endangered. The outline of the distribution on a map forms the extent of occurrence (EOO), a polygon whose area can be measured by the GIS system. The smaller the size of the polygon, the more likely it

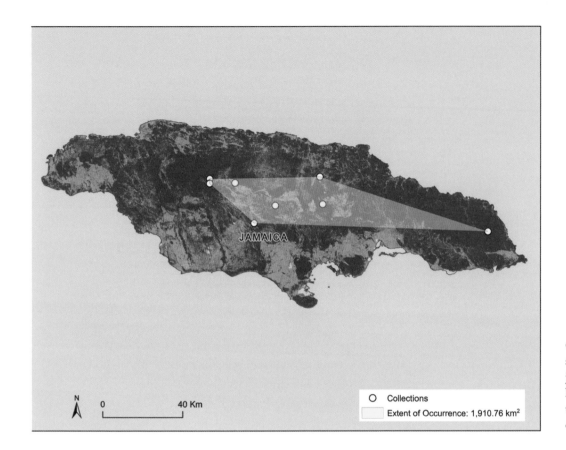

GIS map showing a very small extent of occurrence, just under 2,000 square kilometers, indicating that the species may be at risk of extinction.

is that the species is at risk of extinction. The International Union for Conservation of Nature considers species that cover an area less than 20,000 square kilometers as the most likely to be endangered. Using digitized herbarium specimen data from the world's herbaria to calculate EOO values is a quick way to identify those species with the smallest distributions, the ones that should be prioritized for additional study and possible protection.

With climate change and the fragmentation or absolute loss of natural habitats, many more species will become endangered in the near future. In response, we must increase the rate at which we make new collections of plants and fungi: for many species, now may be our last chance to document them and understand their web of species interactions and role in critical ecosystem services, such as flood mitigation and water purification. We can only hope that this sad reality will lead to a reevaluation of the types of information we gather when we collect a plant, so that we not only preserve a sample of a few stems, leaves, and flowers but also gather data to better understand the light, soil, and water conditions under which the plant grew; the actions of pollinators and seed dispersers that aid in its growth and reproduction; and the fungi that live inside its tissues and break down those tissues when the plant dies. Above all and whenever possible, any collection of living specimens should be accompanied by the collection of seeds or cuttings for cultivation, even if it means a return trip to the site.

Ecological Niche Modeling

For each species of plant and fungus, there is a set of optimal biotic and environmental conditions for its growth and reproduction. These conditions determine the ecological niche of the species. It is difficult to characterize an ecological niche by direct observation, but by overlaying species occurrences obtained from herbarium specimens with a GIS system's available environmental data layers (e.g., land contour and elevation, soil type, temperature ranges, annual rainfall), we can infer key elements of its niche. Then, using modeling algorithms, we can change the environmental factors in the GIS system to match those predicted by climate change projections to see how the distribution of the species may change in response. When this exercise is done with all the species that grow in the area, one can forecast how the whole community may be affected by environmental change—increasing temperature, less rain, higher sea levels, and so forth. The effects suggested by the modeling exercise can then be tested by field or controlled growth experiments.

We can also overlay the distribution patterns of species in a GIS system to find those areas that hold the greatest species diversity, and if we add in information obtained from DNA sequencing of these species, we can begin to understand how that diversity arose, either through species diversification in the area or by introductions from surrounding areas—which in turn can help us decide which areas to protect. Ecosystems in areas with the highest genetic diversity are most likely to remain intact and provide critical services, such as water filtering and oxygen production, as well as spiritual and recreation opportunities.

Phenology

When making herbarium specimens of plants, collectors are particularly concerned with documenting structures that contain diagnostic features needed to distinguish one species from another. Among such features are leaf and flower buds, open flowers, and fruits. These structures also indicate stages of the plant's phenology, or seasonal life events, and thus in addition to using specimens to identify the plant from which they were derived, we can use them to deduce the timing of its annual growth and reproductive cycles. Because we know when and where each specimen

NEW YORK BOTANICAL GARDEN

Gentiana calycosa Griseb.

U.S.A. WYOMING. TETON COUNTY: Teton Range,
Rendezvous Mountain, Cody Basin, south of the Tram
station; 43°35'27"N, 110°53'04"W; T42N R117W S22
m (9765 ft) elevation.

Rocky meadow with Pedicularis groenlandica (which is
flowering) and graminoids.

Corolla dark blue. The margins of the calyx lobes and le
are papillate.

Noel H. Holmgren 15457   19 Augus
Patricia K. Holmgren

was collected, we can compare phenological change over time and space. In 2004 Daniel Primack and colleagues at Boston University published the results of a study that used a variety of information sources, including herbarium specimens, to document the past flowering times of selected plants (in 37 genera) in New England. The study, which compared past and present flowering times and correlated these changes with meteorological data, showed a trend linking increased spring temperatures to earlier flowering. Follow up studies have confirmed that phenological

▲ ▶ Portions of herbarium specimens showing various phenological stages: flower buds, open flowers, and fruits.

trends detected through the study of herbarium specimens correlate with climatological records and field observations, providing yet another way that herbarium specimens, by telling us about past environments, can help us make predictions about the future.

Shifts in phenology, one of the most commonly predicted effects of global change, are a cause for concern because such shifts could lead to a mismatch between plants and other organisms with which plants are interdependent, for example, pollinators. Studies using large sets of digitized herbarium specimens have confirmed that these predicted mismatches are indeed happening. An example is a study by Hutchings et al. (2018) of pollination in the early spider orchid (*Ophrys sphegodes*) in central England. When it flowers, the orchid emits a fragrance that smells like a female buffish mining bee (*Andrena nigroaenea*). The male bee of this species typically emerges in the spring before the female, at the same time that the orchid is flowering. The male bee is attracted to the flower and attempts to copulate with it, picking up pollen from the flower in the process, which it deposits on the next early spider orchid flower that attracts it. A comparative survey of herbarium specimens of *O. sphegodes* over the past 150 years from the Kew and Sloane herbaria; specimens of the buffish mining bee from Oxford's Museum of Natural History; and climate data from central England indicates that warming temperatures are causing the orchids to flower earlier and the female bees to appear earlier. This means that the orchid is now competing with the female bee for the male's attention, and that fewer flowers will be pollinated. Less pollination means fewer offspring for the plant, less genetic diversity, and thus less adaptability to environmental change.

Invasive Species

The process of moving plants from where they grow naturally to new habitats began with the Age of Exploration and has continued ever since. Many plants are moved intentionally for agriculture or horticultural purposes. But there have been many unintentional introductions as well, at a rate that has increased with global traffic. Some inadvertently introduced plants do not persist, but others may become invasive and, through their aggressive growth, disrupt ecosystems by crowding out native plants and their animal pollinators and seed dispersers. Invasive species may also interfere with agriculture, resulting in the loss of billions of dollars annually. Since native species may have highly specific growth requirements, a changing environment favors the growth of invasive species, which usually tolerate a wider range of climatic conditions.

Herbaria can be used to help determine when historical invasions took place and how they spread. When a botanist encounters an unfamiliar species, her first impulse is to collect it. Thus, at least in areas with a long tradition of botanical study, new invasives are likely to be collected soon after they become established, creating an herbarium record of the invasion.

The most perplexing thing about invasive species is that generally the organism is not invasive in its native range—it spreads aggressively only after it is introduced into a new habitat. Perhaps in some cases the organism grows more vigorously in the new location because it is not inhibited by pathogens that it faces at home. It could be that in the new habitat the invader hybridizes with a closely related native species, or with plants of its own species that date from a different invasion event and which therefore differ subtly genetically. Herbarium specimens can be used to document when an invasive species may have entered an area; DNA sequencing of herbarium specimens may show genetic changes that occurred after the invasive reaches a new area, and niche modeling of herbarium data can predict how it may spread. Through such exploration it may be possible to arrive at a general understanding of what causes invasiveness and how to prevent or manage the spread of invasive species.

New Tools for Identification

Despite the centuries of work that have gone into the effort, it can still be difficult to identify plants and fungi, partly because of inadequate training and unavailable or incomplete literature references, but also because an estimated 20% of the world's plants and 50% of fungi have not yet been named. The Global Strategy for Plant

▶ This specimen of *Bromus tectorum* (cheat grass) from the New York Botanical Garden's Steere Herbarium is the earliest known collection of this invasive plant in North America. It is accompanied by a letter dated 13 June 1859 from a Mr. H. Jackson of West Chester, Pennsylvania, to his "respected friend," Alphonso Wood, author of several textbooks on botany, who lived in the Bronx. The purpose of Jackson's letter was to seek Wood's opinion on the specimen, since it did not fit any known local species of the genus.

West Chester, Jan. 13. 1859.

Respected Friend,

The enclosed is a species of Bromus which I have found growing in this place. It seems to agree very well with B ciliatus of Gray but not with B purgans which I perceive is made synonymous with ciliatus and pubescens in thy work. My plant is not the B ciliatus of Darlington's Flora Cestrica. The plant which he has described under this name appears to be the true B pubescens. Dr Darlington informs me that my plant differs from the B ciliatus preserved in his Herbarium. If thou hast duplicate specimens of B purgans I should be much obliged for a specimen. The plant in question is very rare here. It appears to be a stranger and is confined to a single locality. It resembles very closely the B tectorum of Europe, a dried specimen of which I have in my own Herbarium. I should like to have a specimen of Polygala Nuttalli also of Gentiana lutea. Whence can they be obtained? If any Botanist in your section of the country would be willing to make an exchange of

Conservation recognizes the lack of easily accessible tools for identification as the primary impediment to effective conservation measures. Estimates suggest that up to 50% of the still undescribed species of plants may actually already be represented in herbaria but are either misidentified or are in the unidentified backlog that almost every herbarium holds (Bebber et al. 2010).

Machine learning applications such as computer vision and data mining can be applied to digitized herbarium specimens. This work is at an early stage of development but holds great promise as a technology that can build on digitized specimens to broaden the use of herbarium specimens even further. Preliminary attempts to apply machine learning to plant identification from herbarium specimens have shown that algorithms can match images of specimens of unknown identity to a reference set of identified herbarium specimen images with an accuracy rate of 85–90%, and further study of the algorithms and refinement of the processes will likely make the rate of correct matching even higher. If we are able to combine machine learning based on image recognition with data mining of descriptive data from published literature and gene sequence data, we may have a very powerful tool for completing the work started by Luca Ghini and his students in the 16th century.

Machine learning has also been applied to phenological studies. By creating reference sets of specimen images that mark, or mask, the features of interest (e.g., buds, flowers, fruits), algorithms can identify these stages in unmasked specimens, thereby overcoming a very large obstacle in these studies: the time it takes a person to score each specimen for its phenological stage.

▶ The Hand Lens is the public outreach resource of the New York Botanical Garden's C. V. Starr Virtual Herbarium. Posts may tell stories about specimen-based research, collectors, or expeditions, or they may be whimsical, like the Specimen Alphabet: "A" is *Liatris spicata*, collected in Albany County, New York, 1937; "B" is *Wettinia hirsuta*, Panama, 1989; "C" is *Anthurium coripatense*, Bolivia, 1894; "D" is *Cyathea arborea*, Guadeloupe, early 1900s; "E" is *Bactris acanthocarpoides*, French Guiana, 1988; "F" is a caper (*Capparis*), Myanmar, 1948.

Educational Opportunities in Herbaria

Plant blindness is the phenomenon wherein we see plants not as living organisms that are key to our existence but merely as background scenery to our lives. Understanding the importance of plants will help us make the social transformations that will be necessary to reduce our impact on the earth. Data literacy—that is, understanding how big data can be used (and misused) to make decisions that impact our lives—is critical on the job and at the ballot box. Herbaria can play a role in teaching people not only to appreciate plants as living organisms but also in developing the very transferable skills of preparing and analyzing data.

Herbaria have always been used by advanced students doing postgraduate research but traditionally have been inaccessible to younger students. The availability of large quantities of digitized herbarium specimens creates new opportunities to engage younger students, not just to learn about plant biology but also how to

analyze data. For example, a short course offered to high school students through the New York Botanical Garden teaches them to capture data about selected species from herbarium specimens and then to enter these data into a GIS system, in order to determine whether the extent of occurrence falls within the threshold for endangered status. Next, the students evaluate their determination by considering both biological aspects of the species (e.g., life history, pollinators, seed dispersers) and the area's past and predicted land use. This exercise teaches students to evaluate data and understand its limits; they also gain insight into the work involved in science-based conservation, a topic in which most young people have a strong interest. Recognizing the value of collections to create a data-literate future workforce, the National Science Foundation supports Biodiversity Literacy in Undergraduate Education (BLUE; biodiversityliteracy.com), an initiative that is building a publicly available resource for lesson plans that use data from biodiversity collections in a classroom setting.

Human experiences are increasingly virtual in the Internet Age—which can enhance an individual's reaction to seeing rare physical objects, especially historical ones. In nearly 40 years of giving tours of herbaria, I have never met a person who did not express wonder at seeing a specimen collected on one of Cook's voyages, or by Charles Darwin or George Washington Carver. Meaningful examples need not be of major historical figures, however. Specimens collected by a person's great-grandmother or from a beloved childhood haunt may have equally strong effects, as does the aesthetic appeal of the dried specimen itself, or the elegant script of its label or attached correspondence. Herbarium specimens are an engaging entry point for more substantive conversations about plant biology, climate change, and conservation.

Most herbaria do not have dedicated exhibit space, but all have some capacity for showing their specimens through tours or open houses; these can be especially effective when connected to larger public events at the institution—Earth Day celebrations, exhibit openings, parents or alumni weekends, for example. Herbaria can also use their online presence to tell interesting stories about their collections, to demonstrate the value of specimens to particular research projects, to connect herbaria to historical events, or just to show particularly beautiful items. Social media is an ideal venue for images and concise specimen stories; many herbaria share their content through Facebook, Twitter, and Instagram.

Preserving Herbaria

The past decade has been a golden age for herbaria, a time when the value of their holdings has been reintroduced through the application of new technology and techniques. These recent innovations are just the beginning—our specimens collected and preserved through the ages have not yet reached their full potential to serve research and education. However, a lack of appreciation of the value of herbaria, sometimes even by the institutions that house them, makes their future insecure.

▶ The storage room at the New York Botanical Garden's Steere Herbarium. The room is kept cold to prevent insect infestation while specimens wait to be mounted and inserted in the herbarium. The bundles of blue-tagged specimens on the far wall have been identified and are waiting their turn for the mounting room. All other tags mark specimens from expeditions that are awaiting identification. Specimens may be identified by staff or by a network of experts around the world. For some groups and geographical regions, very few experts are available to identify specimens; some specimens stay in this room for many years.

Compared to other types of scientific resources that we need to preserve the well-being of the planet, the basic maintenance needs of an herbarium are low but continual, since herbaria are meant to be preserved in perpetuity. Herbaria require a secure room or rooms with sufficient floor loading for the specimen cabinets, environmental controls to keep the space at a maximum of 70F (21C) and 50% relative humidity, a fire suppression system, and computer network access. Specimen cabinets are the main type of equipment needed in an herbarium. These are most often steel units designed specifically for herbarium specimens, sometimes mounted on a compactor system. Alternatively, some major herbaria store specimens in boxes on shelves. Other equipment needs include freezers for specimen fumigation, computers and microscopes for specimen study, and one or more imaging stations, including camera, computer, and lights for specimen digitization.

◄ Specimens laid out for the public at an herbarium open house. An herbarium room is usually windowless, with banks of cabinets and counter space for laying out the specimens.

► CLOCKWISE,
FROM TOP LEFT
Rows of herbarium cabinets on either side of an aisle. At the Steere Herbarium, the specimen cabinets are mounted on carriages that can be moved with a hand crank to open an aisle between rows of cabinets.

An herbarium cabinet with an open door to show how the specimens are stored on shelves within it at the Steere Herbarium.

Herbarium staff members performing various collections management tasks at the New York Botanical Garden.

The day-to-day work needed to maintain an active herbarium includes preparing newly acquired specimens and filing those specimens; helping onsite users find the specimens they need to study; or pulling specimens from the herbarium to send on loan to another herbarium for study by a researcher there. The herbarium staff, which might range in number from one to 30 individuals depending on the size of the collection, may also identify unnamed specimens or prepare sets of these for an expert in that particular group to identify. The staff may prepare duplicate specimens to send to another herbarium as part of an exchange program, for which they will in turn receive duplicates from that herbarium, which then must be prepared and filed. Sometimes specimens need to be rearranged to reflect the most current thinking about relationships among organisms, or to relieve overcrowding in the cabinets. Older specimens may need repair to reglue specimens or labels, or to replace damaged or non-archival folders or boxes.

Herbarium staff must be constantly on the lookout for evidence of insect pests in the collection—the cigarette beetle (*Lasioderma serricorne*) can obliterate a specimen in a very short time. If an herbarium finds evidence of infestation, it will immediately begin a fumigation regime that may involve freezing or chemical pesticides.

Herbaria that serve their digitized data on their own websites need support for site development and maintenance; the digitized specimen records become their own collection, requiring curation, updating, and management of software, hardware, and networking functions. Most herbaria share their digitized data through one or more data portals, requiring periodic data transfers.

▶ This type specimen of the mushroom *Russula paxilloides*, from C. F. Baker's *Pacific Slope Fungi*, has been nearly destroyed by herbarium beetles.

Rhododendron canadense (L.) Torrey

Det. W. S. Judd & K. A. Kron, FLAS   1988

261
65

INSECT DAMAGE NY

EASTERN FLORA.
Ex. Herb. CARLTON C. CURTIS.

Rhodora Canadensis L.
Thomas N F. N.S.
V-28-
1904

Threats to Herbaria

The shift in focus of biological research in the late 20th century to cellular and subcellular processes in a small number of model organisms meant that more resources were apportioned to research and teaching in those areas, and far less to biodiversity discovery and documentation. This basic research, and the collections that support it, were considered old-fashioned—relics of the age of descriptive biology and not configurable into a hypothesis-driven model of experimental research. The lack of emphasis on biodiversity research and collections propelled a chain reaction wherein funding declined for biodiversity research, resulting in fewer hires of biodiversity specialists, limiting the training opportunities in plant and fungal biodiversity at universities, leading to less use of collections and consequently to diminished space and funding for herbaria, sometimes causing them to fade entirely from institutional consciousness. The new uses for herbarium specimens described earlier have helped to change the direction of this reaction, but we are still in a period where general impressions of value lag behind reality, and many herbaria lack sufficient infrastructure and basic curatorial staff.

In the United States, the post–World War II building boom created many of the facilities in which herbaria were originally housed. Those buildings, now more than half a century old, are often in need of upgrades or in some cases must be demolished. New or remodeled buildings often give less space to resources like herbaria. Sometimes collections are moved to basements, where they are at risk of flooding, or to attic areas with inadequate climate control. Large collections take up a significant amount of space, and as land becomes scarcer in urban areas, where most collections are concentrated, the only option may be to move them to remote satellite locations, where they may become isolated from their host institution, programmatically as well as physically.

Unable or unwilling to meet the infrastructure and staffing needs, some institutions have decided to divest themselves of their herbaria. In a recent case that received a great deal of press coverage, the University of Louisiana at Monroe presented their extensive herbarium and fish collections with an ultimatum: move from their current facility or be discarded. The building that housed the collections was slated to be demolished to make way for a new football stadium. Fortunately, a new home for the herbarium was found at the Botanical Research Institute of Texas and for the fish at the Tulane University Natural History Museum.

Transfer of herbaria from one institution to another has happened throughout history, and this action maintains (sometimes even improves) the care and accessibility of the specimens. However, concentrating our specimen heritage in

◄ Specimen of the flowers of *Rhododendron canadense*; those on the left have been eaten by beetles.

an increasingly limited number of institutions has risks as well. One is the danger of loss or damage through weather or human-mediated disasters, as with the loss of Berlin's herbarium in World War II. The substantial National Herbarium in Caracas, Venezuela, is currently experiencing neglect from lack of staff and looting of air conditioners and other critical infrastructure. Even large institutions are not safe from administrative decisions to compromise collections care—in fact, several herbaria with more than a million specimens have only one paid staff member! Also, each herbarium that we lose through consolidation diminishes the strength of our community, and restricts access to the plant and fungal history of that area to the people who live in that area, where that information is most needed.

The question of what we can do to keep herbaria and all natural history collections strong and able to address the environmental challenges we face is foremost in the minds of collections managers. Individually or through organizations such as the Society for the Preservation of Natural History Collections and the Society of Herbarium Curators, we are working to broaden the understanding of the relevance of herbaria to life in the 21st century. Recognizing the need to broaden the base of support for collections, the National Science Foundation has commissioned a study by the National Academies of Science to recommend strategies to safeguard and sustain these irreplaceable research and educational resources.

How You Can Help

There are many ways that you can help support the world's herbaria and thereby contribute to finding solutions to the environmental challenges we face. Almost every herbarium welcomes volunteers, and in fact many herbaria depend on them for day-to-day herbarium maintenance. In return, volunteers often receive membership benefits at museums and botanic gardens and have the opportunity to participate in field trips and identification courses. *Index Herbariorum* (sweetgum.nybg.org/science/ih) can help you find the herbarium nearest to where you live. An email inquiry about volunteering to the correspondent listed for that herbarium will very likely be met with an enthusiastic response.

If you want to volunteer but are unable to come to an herbarium, online opportunities are available as well. Using such platforms as Notes from Nature (notesfrom-nature.org), citizens can participate in the digitization of natural history collections by transcribing specimen labels from images into web forms. Volunteers have already transcribed data from more than half a million specimens of plants, fungi, insects, and birds! A four-day campaign held each October, Worldwide Engagement for

Digitizing Biocollections, or WeDigBio (wedigbio.org), aims to engage as many participants as possible in digitizing natural history collections. Museums, herbaria, universities, and other institutions host onsite as well as online events on these days where, in addition to transcribing data, participants meet likeminded people in person and virtually by connection to other transcription events around the world. Collections tours, demonstrations, and even short field trips are often offered to transcribers as well. Often more than 50,000 specimen records are transcribed by volunteers during a WeDigBio event.

Volunteers report that in addition to the knowledge and satisfaction they gain from providing support to herbaria, this work also helps them cope with the sense of helplessness they sometimes feel about the future—while individually powerless to stop environmental degradation or species loss, knowing that they may be contributing to solutions, even in a small way, provides some comfort and restores some optimism about the future.

Throughout this book are examples of plant collectors who were not formally trained as scientists but who have made tremendous contributions to our knowledge of plant and fungal biodiversity. Today, just as in the past, citizens continue to make collections that are highly valued by herbaria and to publish articles and books of the highest academic standards. Probably the easiest way to get started is with an application such as iNaturalist (inaturalist.org) or, for fungi, Mushroom Observer (mushroomobserver.org). With these applications, you can record observations of plants and fungi, and often receive accurate identifications for the species you photograph. Observations are not useful for research in all the ways that actual specimens are, but they do document phenological events and the spread of invasives, for example, and data from these applications are increasingly being incorporated into research projects. If you have an interest in making scientific collections, this is a great way to make a contribution that benefits research worldwide. We must continue to document plants and fungi as we enter a critical period of environmental change, and there are not enough appropriately trained professional botanists to do this work.

Developing the techniques for making scientifically valuable collections does take some training and practice, and one must be aware of the laws that govern the collection and shipment of specimens. So where does one learn how to make collections? Your local herbarium may be able to point you to botanic gardens, field stations, naturalist clubs, universities, and other resources that offer the general public such learning opportunities. Maintaining a personal herbarium may be difficult because of the environmental conditions needed in order to prevent mold and insects, but established herbaria will usually welcome the addition of specimens that are ethically and legally collected, well preserved, and documented.

The North American Mycoflora Project is a citizen science initiative to document the fungi of the continent through herbarium specimens and DNA sequence data. For many years, citizens interested in fungi have come together to collect fungi and learn to identify them, often under the auspices of the North American Mycological Association (namyco.org) or one of its affiliated clubs. In many cases people begin their interest in fungi in order to identify wild mushrooms for food, but often this interest grows far beyond collecting for the table, and many citizen mycologists have become internationally recognized experts in the taxonomy of particular groups. The North American Mycoflora Project harnesses the collective power of the many people who love to explore the woods for new and unusual fungi and provides a mechanism for them to share the results of this beloved leisure activity with the rest of the world through specimens deposited in herbaria and gene sequences deposited in GenBank.

Keeping herbaria supported for future generations will not be easy, but as many of the stories in the book have indicated, building and maintaining herbaria has always been challenging! Many of our herbarium heroes suffered greatly to help build this heritage of plant knowledge: Jeanne Baret, who endured the difficulties of fieldwork while disguised as a man and even more trials after her identity was discovered; Richard Spruce, who ruined his health documenting Amazonian and Andean plants; and Alice Eastwood, who lost everything she owned to save the California Academy of Sciences type specimens. We owe it to these heroes, to ourselves, and to future generations to continue this tradition, to help ensure a future in which plants and fungi continue not only to provide the fundamental needs for our existence but also to delight our senses and imagination.

Acknowledgments

I have many people to thank for their contributions to this book. First of all, I thank my editors at Timber Press, Tom Fischer and Franni Farrell, for their encouragement, expertise, and excellent judgment. Their suggestions vastly improved this book.

The institutions and staff that shared images and information include the Arnold Arboretum, Larissa Glasser, Library Assistant; the Bavarian State Library (Bayerische Staatsbibliothek); California Academy of Sciences, Rebekah Kim, Head Librarian, and Debra Trock, Director of Science Collections; Cornell University, Kathie Hodge, Professor and Director of the Cornell Plant Pathology Herbarium; Drexel University and the Academy of Natural Sciences, Rick McCourt, Director, Center for Systematic Biology and Evolution, and Daniel Thomas, Collection Manager; Gerald Ford Library, Jeffrey Senger, Archives Technician; Harvard University, Danielle Castronovo, Archivist, Gray Herbarium Library and Archives, and Donald Pfister, Professor; Missouri Botanical Garden, Doug Holland, Library Director; National Herbarium, Pretoria, Erich Van Wyk, Lufuno Makwarela, and M. S. Mothogane; Natural History Museum, Denmark, Olof Ryding; Natural History Museum, London, Curators Mark Carine and Fred Rumsey; Natural History Museum, Paris, Marc Pignal, Curator; Oxford University Herbaria, Stephen Harris, Druce Curator; Princeton University Library, AnnaLee Pauls, Reference Assistant; Royal Botanic Garden, Madrid, Eva García Ibáñez, Herbarium Technician, and Abel Blanco Asenjo, Archivist; Royal Botanic Gardens, Kew, Pei Chu, Publishing Assistant; Royal Botanic Gardens Victoria, Pina Milne, Collections Manager, and Angharad Johnson, Digitising Officer; Smithsonian Institution, Heidi Stover, Archivist, and Ingrid Lin, Multimedia Manager; Sutro Library, Mattie Taormina, Archivist; University of California, Berkeley, and Jepson Herbarium, Amy Kasameyer, Archivist, Kim Kersh, Curatorial Associate, and Brent Mishler, Director of the Herbarium; University of California Museum of Paleontology, Trish Roque, Web Manager; University of Cape Town, Bolus Herbarium, Cornelia Klak, Curator;

University of Kentucky, Special Collections Research Libraries, Daniel Weddington, Research Services Archivist; and Western Australian Museum, Jeremy Green, Head, Department of Maritime Archaeology.

For the use of their personal photographs, I thank Roy Halling, James Lendemer, and Amy Weiss of the New York Botanical Garden, Barry Rice of Sierra College, and Michael Wood of Mykoweb. I very much appreciated the chance to visit several collections that were critical to my understanding of herbarium history. Isabelle Charmantier, Head of Collections, was my guide to the Linnean Herbarium at the Linnean Society in London; Christel Schollaardt, Collection Manager, Tinde van Andel, Senior Researcher, and Anastasia Stefanaki, Guest Researcher, showed and explained to me both the Rauwolf and *En Tibi* herbaria at the Naturalis Biodiversity Center; and Mark Carine, Curator, did likewise for critical early specimens held in the Sloane Herbarium at the Natural History Museum in London.

Several of my colleagues graciously read various chapters of my book. For this kind assistance I thank John Clarkson, Principal Botanist, Queensland Parks and Wildlife Service; Peter Jobson, Senior Botanist, Northern Territories Department of Land Resources Management; Marianne Le Roux, E-Flora Coordinator, National Herbarium, Pretoria; Lucía Lohmann, University of São Paulo; Muthama Muasya, Head, Department of Biological Sciences, University of Cape Town; Peter Raven, President Emeritus, Missouri Botanical Garden; and Roy Halling, Emeritus Curator, and William Wayt Thomas, Curator, New York Botanical Garden.

I am very grateful to Elizabeth Gjieli, Manager, New York Botanical Garden GIS Laboratory, for the maps she cheerfully and skillfully produced for this book, and to my colleagues Susan Fraser, Director, Stephen Sinon, Archivist, and Sukshma Dittakavi, Digitizer, of the LuEsther T. Mertz Library of the New York Botanical Garden for helping me obtain critical references and illustrations from this wonderful collection. I thank the management staff of the William and Lynda Steere Herbarium for their support and assistance throughout the researching and writing of this book, and for giving me the dream job of serving as their director: Ginger Apolo, Curatorial Assistant; Laura Briscoe, Collection Manager; Elizabeth Gjieli, GIS Laboratory Manager; Leanna McMillin, Curatorial Assistant; Matthew Pace, Assistant Curator; Joel Ramirez, Web Developer; Edgardo Rivera, Curatorial Assistant; Nicole Tarnowsky, Assistant Director; Melissa Tulig, Director of Biodiversity Information; Kimberly Watson, Digital Asset Manager; Amy Weiss, Collection Manager; Charles Zimmerman, Herbarium Collections and Outreach Administrator. Many other colleagues at the New York Botanical Garden contributed to this work by listening to the seminars I gave on each chapter and offering questions and perspectives. Former New York Botanical Garden President Gregory Long gave much-needed encouragement in the formative stages of this project, and I very much appreciate his active interest in the project.

I thank Chrissie and David and my fellow yogis at The Stretch yoga studio in Yonkers, New York, for their fellowship during the writing of this book—many a seemingly impossible writing obstacle seemed much less intractable after a practice with this supportive and fun community! I thank my daughter Mary and son-in-law Alberto for their love, encouragement, and technical advice, and the pleasure of their company from time to time to distract me from the book. Most of all, I thank my steadfast and some might say long-suffering husband Roy, for his never-wavering love, support, and wisdom, for a fresh rose in the kitchen windowsill whenever possible, and for a cup of mint tea at bedtime.

01128485

Didymodon ~~styrideus~~ sp. n. *Jur.*
    styriacus

1878

EXAMINED BY RICHARD H. ZANDER

Leptodontium flexifolium (With.) Hamp
    24 Aug 19 66

Didym styriacus Jur sp nov R
    aff. D. flexif.
        Styria

0112

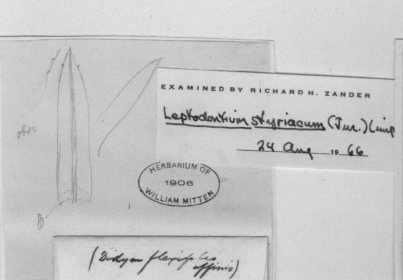

EXAMINED BY RICHARD H. ZANDER

Leptodontium styriacum (Jur.) Limp
    24 Aug 19 66

(Didym flexifolio
    affinis)

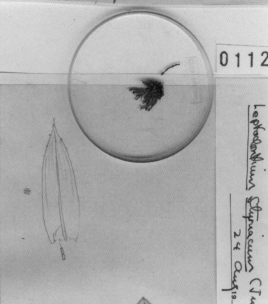

Leptodontium styriacum (Jur.)
24 Aug 19 6

Select References

Anderson, J. G. T. 2013. *Deep Things Out of Darkness: A History of Natural History*. Berkeley: University of California Press.

Arber, A. 1938. *Herbals: Their Origin and Evolution*. Cambridge: Cambridge University Press.

Bebber, D. P., et al. 2010. "Herbaria are a major frontier for species discovery." *Proceedings of the National Academy of Sciences of the United States of America* 107:22169–22171.

Beerling, D. J., and W. G. Chaloner. 1993. "Stomatal density responses of Egyptian *Olea europaea* L. leaves to CO_2 change since 1327 BC." *Annals of Botany* 71:431–435.

Berkeley, E., and D. Smith Berkeley. 1963. *John Clayton: Pioneer of American Botany*. Chapel Hill: University of North Carolina Press.

Blunt, W. 2004. *Linnaeus: The Compleat Naturalist*. New York: Viking Press.

Bonta, M. M. 1991. *Women in the Field: America's Pioneering Women Naturalists*. College Station: Texas A&M University Press.

Clarke, P. A. 2008. *Aboriginal Plant Collectors: Botanists and Australian Aboriginal People in the Nineteenth Century*. Kenthurst, New South Wales: Rosenberg Publishing.

Cox, E. H. M. 1945. *Plant-Hunting in China. A History of Botanical Exploration in China and the Tibetan Marches*. London: Collins.

Cutright, P. R. 2003. *Lewis and Clark: Pioneering Naturalists*. 2d ed. Lincoln: University of Nebraska Press.

Dash, M. 1999. *Tulipomania: The Story of the World's Most Coveted Flower and the Extraordinary Passions It Aroused*. London: Victor Gollancz.

Dupree, H. 1959. *Asa Gray*. Cambridge: Belknap.

Ewan, J. 1950. *Rocky Mountain Naturalists*. Denver: University of Denver Press.

Eyles, V. A. 1973. *Introduction to John Woodward's Brief Instructions for Making Observations in All Parts of the World, 1696*. Sherborn Fund Facsimile No. 4. London: Society for the Bibliography of Natural History.

Fang, R., et al. 2018. "Country Focus: China." In *State of the World's Fungi*. Royal Botanic Gardens, Kew.

George, A. S. 2009. *Australian Botanist's Companion*. Kardinya, Western Australia: Four Gables Press.

Grove, R. H. 1995. *Green Imperialism: Colonial Expansion, Tropical Island Edens and the Origins of Environmentalism 1600–1860*. Cambridge: Cambridge University Press.

Hutchings, M. J., et al. 2018. "Vulnerability of a specialized pollination mechanism to climate change revealed by a 356-year analysis." *Botanical Journal of the Linnean Society* 186:498–509.

Johnson, V. 2018. *American Eden: David Hosack, Botany, and Medicine in the Garden of the Early Republic*. New York: Liveright.

McLauchlan, K. K., et al. 2010. "Thirteen decades of foliar isotopes indicate declining nitrogen availability in central North American grasslands." *New Phytologist* 187:1135–1145.

Miller, D. P., and P. H. Reill, eds. 1996. *Visions of Empire*. Cambridge: Cambridge University Press.

Morel, R. S. 2013. "James Cunningham." *Untold Lives*. British Library. blogs.bl.uk/untoldlives/2013/05/james-cunningham-the-unluckiest-botanist-in-asia.html.

Mori, S. 2013. "Flora Brasiliensis." *NYBG Plant Talk*. nybg.org/blogs/plant-talk/2013/05/science/flora-brasiliensis-how-a-19th-century-flora-continues-to-inspire.

Morton, A. G. 1981. *History of Botanical Science*. New York: Academic Press.

Moyal, A. 1986. *A Bright and Savage Land*. Ringwood, Victoria: Penguin Books.

Needham, J., G. Lu, and H. Huang. 1986. "Biology and Biological Technology, Part 1 Botany." In *Science and Civilisation in China*, vol. 6. Edited by J. Needham. Cambridge: Cambridge University Press.

Nepi, C. 2016. "Botanical Collecting, Herbaria, and the Understanding of Nature." In *Changing Perceptions of Nature*. Edited by I. Convery and P. Davis. Newcastle University.

Olsen, P. 2013. *Collecting Ladies: Ferdinand von Mueller and Women Botanical Artists*. National Library Australia.

Preston, D., and M. Preston. 2004. *A Pirate of Exquisite Mind: The Life of William Dampier*. New York: Walker & Co.

Primack, D., et al. 2004. "Herbarium specimens demonstrate earlier flowering times in response to warming in Boston." *American Journal of Botany* 91:1260–1264.

Ridley, Glynis. 2011. *The Discovery of Jeanne Baret: A Story of Science, the High Seas, and the First Woman to Circumnavigate the Globe*. New York: Broadway Books.

Riffenburgh, B. 2017. *The Great Explorers and Their Journeys of Discovery*. London: André Deutsch.

Robertson, J. 1990. *The Magnificent Mountain Women: Adventures in the Colorado Rockies*. Lincoln: University of Nebraska Press.

Rudin, S. M., et al. 2017. "Retrospective analysis of heavy metal contamination in Rhode Island based on old and new herbarium specimens." *Applications in Plant Sciences* 5:1600108.

Saint-Lager, J. P. 1885. *Histoire des Herbiers*. Paris: J. B. Baillière.

Sawidis, T., et al. 1997. "Cesium-137 monitoring using lichens from Macedonia, northern Greece." *Canadian Journal of Botany* 75:2216–2223.

Sumner, R. 1993. *A Woman in the Wilderness: The Story of Amalie Dietrich in Australia*. Kensington, Australia: New South Wales Press.

United Nations Environment. 2019. Global Strategy for Plant Conservation. cbd.int/gspc/targets.shtml.

Victor, J. E., et al. 2015. *Strategy for Plant Taxonomic Research in South Africa 2015–2020*. SANBI Biodiversity Series 26. Pretoria: South African National Biodiversity Institute.

Williams, G. 2013. *Naturalists at Sea*. New Haven: Yale University Press.

Woodward, F. I. 1987. "Stomatal numbers are sensitive to increases in CO_2 from pre-industrial levels." *Nature* 327:617–618.

World Research Foundation. 2017. The Badianus Manuscript. wrf.org/ancient-medicine/badianus-manuscript-americas-earliest-medical-book.php.

Wulf, A. 2016. *The Invention of Nature: Alexander von Humboldt's New World*. New York: Alfred Knopf.

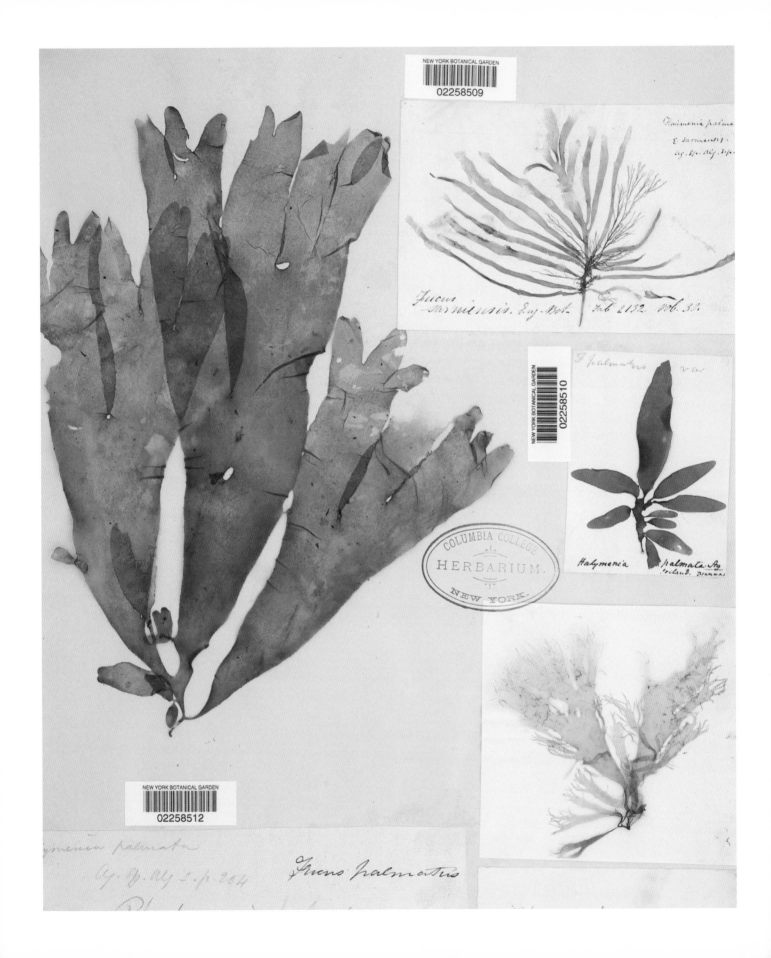

Halimenia palmata

E. sarniensis.

Ag. Sp. Alg. 1.p.

Fucus

sarniensis. Eng. Bot. Tab 2132. Vol. 30.

F. palmatus var

Halymenia palmata Ag.

Ireland. Drummas

ymenia palmata

Ag. Sp. Alg 1. p. 204

Fucus palmatus

Select Herbaria

Ordered alphabetically by their official, internationally recognized abbreviation. Please check Index Herbariorum (sweetgum.nybg.org/science/ih) for the most up-to-date information about these and other herbaria.

AA (see GH)

AD
State Herbarium of South Australia
Adelaide Botanic Garden
Adelaide, Australia
Founded 1954; 1,040,000 specimens
flora.sa.gov.au

AMES (see GH)

AMWH
Amway Herbarium
Amway Corporation
Ada, Michigan, U.S.A.
Founded 2018; 10 specimens

B
Herbarium
Botanical Garden and Botanical Museum Berlin-Dahlem
Berlin, Germany
Founded 1815; 3,800,000 specimens
bgbm.org

BAA
Gaspar Xuarez Herbarium
University of Buenos Aires
Buenos Aires, Argentina
Founded 1962; 400,000 specimens
agro.uba.ar/catedras/botanica_sistematica/herbario

BM
Sloane Herbarium
Natural History Museum
London, England
Founded 1753; 5,200,000 specimens
nhm.ac.uk

BOL
Bolus Herbarium
University of Cape Town
Cape Town, South Africa
Founded 1865; 373,000 specimens
bolus.uct.ac.za

BOLO
Botanical Garden and Herbarium
University of Bologna
Bologna, Italy
Founded 1570; 130,000 specimens
sma.unibo.it/it/il-sistema-museale/
orto-botanico-ed-erbario/collezioni

BPI
U.S. National Fungus Collections
USDA Agricultural Research Service
Beltsville, Maryland, U.S.A.
Founded 1869; 950,000 specimens
nt.ars-grin.gov/fungaldatabases/specimens/
specimens.cfm

BR
Herbarium
Meise Botanic Garden
Meise, Belgium
Founded 1870; 4,000,000 specimens
plantentuinmeise.be

BRI
Queensland Herbarium
Department of Environment and Science
Brisbane, Australia
Founded 1881; 871,000 specimens
qld.gov.au/environment/plants-animals/plants/herbarium

BRIT
Philecology Herbarium
Botanical Research Institute of Texas
Fort Worth, Texas, U.S.A.
Founded 1987; 1,482,000 specimens
brit.org/herbarium

CAS
Herbarium
California Academy of Sciences
San Francisco, California, U.S.A.
Founded 1853; 2,300,000 specimens
calacademy.org/scientists/botany

CGE
Herbarium
Cambridge University
Cambridge, England
Founded 1761; 1,100,000 specimens
cambridgeherbarium.org

CHARL
Herbarium
Charleston Museum
Charleston, South Carolina, U.S.A.
Founded 1773; 25,000 specimens

CINC
Herbarium
University of Cincinnati
Cincinnati, Ohio, U.S.A.
Founded 1925; 125,000 specimens

CUP
Plant Pathology Herbarium
Cornell University
Ithaca, New York, U.S.A.
Founded 1907; 400,000 specimens
plantpath.cornell.edu/CUPpages

DBN
Augustine Henry Herbarium
National Botanic Gardens
Glasnevin, Ireland
Founded 1847; 600,000 specimens

DOV
Claude E. Phillips Herbarium
Delaware State University
Dover, Delaware, U.S.A.
Founded 1977; 150,000 specimens
herbarium.desu.edu

E
Herbarium
Royal Botanic Garden Edinburgh
Edinburgh, Scotland
Founded 1839; 3,000,000 specimens
rbge.org.uk

ECON (see GH)

F
Herbarium
Field Museum
Chicago, Illinois, U.S.A.
Founded 1893; 2,700,000 specimens
fieldmuseum.org/explore/department/botany/collections

FH (see GH)

FI
Herbarium Universitatis Florentinae
Natural History Museum
Florence, Italy
Founded 1842; 5,000,000 specimens
parlatore.msn.unifi.it/hci_italy_web.html

G
Herbarium
Conservatory and Botanical Garden of Geneva
Geneva, Switzerland
Founded 1824; 6,000,000 specimens
ville-ge.ch/cjb

GH
Harvard University Herbaria
Harvard University
Cambridge, Massachusetts, U.S.A.
Founded after the arrival of Asa Gray in 1842;
5,000,000 specimens; includes, besides the Gray
Herbarium (GH), the Arnold Arboretum (AA), Oakes
Ames Orchid Herbarium (AMES), Economic Botany
Herbarium (ECON), Farlow Herbarium (FH), and
New England Botanical Club (NEBC)
huh.harvard.edu

HO
Tasmanian Herbarium
Tasmanian Museum and Art Gallery
Sandy Bay, Australia
Founded 1928; 300,000 specimens

HUDC
Charles S. Parker Herbarium
Howard University
Washington, D.C., U.S.A.
Founded 1915; 12,000 specimens
midatlanticherbaria.org

IBSC
Herbarium
Chinese Academy of Sciences
South China Botanical Garden
Guangzhou, P.R.C.
Founded 1928; 1,000,000 specimens

JBB
Herbarium
José Celestino Mutis Botanical Garden
Bogotá, Colombia
Founded 1985; 21,400 specimens
colecciones.jbb.gov.co/herbario

JEPS (see UC)

K
Herbarium
Royal Botanic Gardens, Kew
Richmond, England
Founded 1852; 8,125,000 specimens
kew.org/science/collections

L
National Herbarium
Naturalis Biodiversity Center
Leiden, Netherlands
Founded 1829; 5,000,000 specimens
naturalis.nl/en/science

LINN
Herbarium
Linnean Society of London
London, England
Founded 1730; 33,800 specimens
linnean-online.org

LISU
Herbarium
Botanical Garden and National Museum
of Natural History
Lisbon, Portugal
Founded 1839; 300,000 specimens

MA
Herbarium
Royal Botanic Garden
Madrid, Spain
Founded 1755; 1,160,000 specimens
rjb.csic.es

MANCH
Herbarium
Manchester Museum, University of Manchester
Manchester, England
Founded 1860; 1,000,000 specimens
museum.manchester.ac.uk/collection/plants/

MC
Avery Island Archives
McIlhenny Company
Avery Island, Louisiana, U.S.A.
Founded 2016; 258 specimens

MEL
National Herbarium of Victoria
Royal Botanic Gardens Victoria
Melbourne, Australia
Founded 1853; 1,500,000 specimens
rbg.vic.gov.au

MELU
Herbarium
University of Melbourne
Parkville, Australia
Founded 1926; 150,000 specimens
biosciences.unimelb.edu.au/engage/
the-university-of-melbourne-herbarium

MO
Herbarium
Missouri Botanical Garden
St. Louis, Missouri, U.S.A.
Founded 1859; 6,850,000 specimens
missouribotanicalgarden.org

N
Herbarium
Nanjing University
Nanjing, P.R.C.
Founded 1915; 150,000 specimens

NAVA
Navajo Nation Herbarium
Northern Arizona University
Flagstaff, Arizona, U.S.A.
Founded 1991; 13,000 specimens
nndfw.org/nnhp/nava.htm

NBG
Compton Herbarium
South African National Biodiversity Institute
Claremont, South Africa
Founded 1937; 507,000 specimens
sanbi.org/biodiversity/foundations/
biosystematics-collections/compton-herbarium

NCATG
Herbarium
North Carolina A&T State University
Greensboro, North Carolina, U.S.A.
Founded 1977; 10,000 specimens

NEBC (see GH)

NEZ
Nez Perce Tribe Herbarium
Wildlife Division, Nez Perce Tribe
Lapwai, Idaho, U.S.A.
Founded 1990; 1,300 specimens

NH
KwaZulu-Natal Herbarium
South African National Biodiversity Institute
Durban, South Africa
Founded 1882; 110,000 specimens

NSW
National Herbarium of New South Wales
Royal Botanic Garden Sydney
Sydney, Australia
Founded 1896; 1,425,000 specimens
rbgsyd.nsw.gov.au

NY
William and Lynda Steere Herbarium
New York Botanical Garden
Bronx, New York, U.S.A.
Founded 1891; 7,921,000
nybg.org

OXF
Fielding-Druce Herbarium
University of Oxford
Oxford, England
Founded 1621; 500,000 specimens
herbaria.plants.ox.ac.uk/bol/oxford

P
National Herbarium
Natural History Museum
Paris, France
Founded 1635; 6,000,000 specimens
science.mnhn.fr/institution/mnhn/collection/p/item/
search/form

PC
Cryptogamic Herbarium
Natural History Museum
Paris, France
Founded 1904; 2,000,000 specimens

PE
Chinese National Herbarium
Institute of Botany, Chinese Academy of Sciences
Beijing, P.R.C.
Founded 1928; 2,650,000 specimens
pe.ibcas.ac.cn

PERTH
Western Australian Herbarium
Perth, Australia
Founded 1929; 807,000 specimens
dpaw.wa.gov.au/plants-and-animals/wa-herbarium

PH
Herbarium
Academy of Natural Sciences of Drexel University
Philadelphia, Pennsylvania, U.S.A.
Founded 1812; 1,430,000 specimens
ansp.org/research/systematics-evolution/botany

PR
Herbarium
National Museum
Prague, Czech Republic
Founded 1818; 2,000,000 specimens

PRE
National Herbarium
South African National Biodiversity Institute
Pretoria, South Africa
Founded 1903; 1,200,000 specimens
sanbi.org/programmes/biosystematics/
national-herbarium

PREM
National Collection of Fungi
Plant Protection Research Institute
Pretoria, South Africa
Founded 1905; 60,000 specimens

R
Herbarium
National Museum
Rio de Janeiro, Brazil
Founded 1831; 600,000 specimens
museunacional.ufrj.br

RB
Dimitri Sucre Herbarium
Rio de Janeiro Botanical Garden
Rio de Janeiro, Brazil
Founded 1890; 800,000 specimens
jbrj.gov.br/colecoes/biologicas

S
Herbarium
Swedish Museum of Natural History
Stockholm, Sweden
Founded 1739; 4,570,000 specimens
nrm.se/english/researchandcollections/botany/
collections.565_en.html

TCD
Herbarium
Trinity College
Dublin, Ireland
Founded 1835; 300,000 specimens
tcd.ie/Botany/herbarium

UC and JEPS
University and Jepson Herbaria
University of California
Berkeley, California, U.S.A.
Founded 1872; 2,100,000 specimens (US),
96,000 specimens (JEPS)
ucjeps.berkeley.edu

UPS
Herbarium
Museum of Evolution, Uppsala
University
Uppsala, Sweden
Founded 1785; 3,100,000
130.238.83.220:81/home.php

US
U.S. National Herbarium
Smithsonian Institution
Washington, D.C., U.S.A.
Founded 1848; 5,100,000 specimens
naturalhistory.si.edu/botany

VEN
National Herbarium
Central University of Venezuela
Caracas, Venezuela
Founded 1921; 450,000 specimens

13525

13525

THE NEW YORK BOTANICAL GARDEN
Plants of Hispaniola
Dominican Republic

Photo Credits

Index

Brazilian National Research Council, 188
brazilwood, 177, 178
Brewer, William H., 139
Brief Instructions for Making Observations . . .
	(Woodward), 41
Britton, Elizabeth, 153, 156–157, 161
Britton, Nathaniel, 140, 153, 156–157
Broken Bay, 168
Bromus tectorum, 230
Brooks, Sarah, 172
Brown, Robert, 44, 168
brown alga, 150, 201
bryophytes, 25–26, 27, 28, 93, 100, 145, 147, 159, 162,
	169, 189, 212
Bungaree, 168
Burchell, William J., 207–208
Burnett, Mr., 82–83

Cactaceae of the Boundary (Engelmann), 134
cactus, 134
Caesalpinia, 18
Caesalpinia pulcherrima, 19
Caldas, Francisco José de, 86
California, 118, 153, 154
California Academy of Sciences, 137, 139, 143–145,
	146
California Geological Survey, 137, 139
California pitcherplant, 154
Canada, 62, 107, 118
Canary Islands, 51
Canhos, Dora, 191
Cape Horn, 51, 71
Cape of Good Hope, 47, 51, 92, 206
caper, 232
Cape Verde Islands, 51
Capparis, 232
carbon dioxide (CO$_2$), 224–225
Caribbean, 47, 104, 157, 178
Carnegiea gigantea, 134
Carver, George Washington, 152–153, 234

Casa de Botánica, 86, 88
Cassone, Felice, 177
Catalog of Plants and Fungi of Brazil, 189
Catcheside, David, 174
Catesby, Mark, 104
Cattleya elongata, 186
Ceanothus jepsonii, 147
Centro de Referencia em Informacao Ambiental
	(CRIA), 191
Cephalorhynchus commersonii, 78
Cereus giganteus, 134
Cesalpino, Andrea, 18, 19
Ceylon, 96
Chamaedoris peniculum, 165
Charaka Samhita, 31
Charles III, King of Spain, 84
Charleston Museum Herbarium, 109
cheat grass, 230
Chen Huan-Yong, 195, 197, 202–203
Chernobyl reactor accident, 224
Chile, 84, 93, 153, 159
China. *See* People's Republic of China
Chinese Academy of Sciences (CAS), 201, 202
Chinese Academy of Traditional Medicine, 32
Chinese Economic Trees (Chen), 195
Chinese fir, 192
Chinese medicine, 193
Chinese Virtual Herbarium, 204
Christina, Queen of Sweden, 22
Chrysomyxa bambusae, 201
cigarette beetle, 236
Cinchona, 89, 96
Cinchona officinalis, 90
Cinnamomum verum, 40
cinnamon, 40
citizen scientists, 241–242
Cladonia verticillata, 27
Clark, William, 113–117, 124, 156
Clarkia, 115
Clarkia pulchella, 117